Lecture Notes in Mathematics

Edited by A. Dold and B. Eckmann

736

Ehrhard Behrends

M-Structure and the Banach-Stone Theorem

Springer-Verlag
Berlin Heidelberg New York 1979

Author

Ehrhard Behrends
I. Mathematisches Institut
der Freien Universität
Hüttenweg 9
D-1000 Berlin 33

AMS Subject Classifications (1978): 46B20, 46B25, 46E15

ISBN 3-540-09533-0 Springer-Verlag Berlin Heidelberg New York
ISBN 0-387-09533-0 Springer-Verlag New York Heidelberg Berlin

Library of Congress Cataloging in Publication Data
Behrends, Ehrhard, 1946-
M-structure and the Banach-Stone theorem.
(Lecture notes in mathematics ; 736)
Bibliography: p.
Includes indexes.
1. Banach spaces--M-structure. 2. Banach Stone theorem. I. Title. II. Series: Lecture notes in mathematics (Berlin) ; 736.
QA3.L28 no. 736 [QA322.2] 510'.8s [515.73] 79-19677
ISBN 0-387-09533-0

Printing and binding: Beltz Offsetdruck, Hemsbach/Bergstr.
2141/3140-543210

Introduction

The mathematical study of M-structure theory has been going on since the middle of the sixties. No comprehensive account has been published; the present notes are an attempt to fill this gap.
The aim is twofold: to make available to potential users the most important results of M-structure theory and to present some applications of this theory to generalizations of the Banach-Stone theorem.

In the hope of attracting some new people to do research on these subjects the author has attempted to make the contents understandable to those who have not yet worked in this field.
The notes are *self-contained* (with the possible exception of the discussion of some examples), and the proofs are *detailed* and *elementary*. The reader is only assumed to be familiar with the fundamental results of functional analysis (Hahn-Banach theorem, Alaoglu-Bourbaki theorem, Krein-Milman theorem, Krein-Šmulian theorem), topology (compactifications, partitions of unity), and Boolean algebras (the Stone representation theorem).
The text contains many examples, and the definitions and proofs are illustrated by a number of figures.

M-structure theory

Roughly speaking, M-structure theory measures to what extent a given Banach space X behaves like a CK-space (= a space of continuous functions on a compact Hausdorff space)[1]. This is done by defining certain operators and subspaces (the "M-structure" of X: the operators in the centralizer, the M-summands, and the M-ideals) in such a way that CK-spaces have sufficiently many of these operators and subspaces.
The essential features of the theory have been developed by Cunning-

[1] This motivates the name "M-structure theory" (recall that the CK-spaces are just the M-spaces with order unit).

ham [30] and Alfsen and Effros [3],[4]. Cunningham used the concept of centralizer ("the maximal M-structure") to determine, for a given Banach space X, the maximal compact space such that X is a CK-module of a particular type. The definition of "M-ideal" as well as a number of important results and applications of M-structure theory to different branches of functional analysis are due to Alfsen and Effros. Further results and applications have been contributed by several authors (cf. the references in chapter 6).
Chapters 1 - 6 of these notes are devoted to a more or less systematic development of the concepts of M-structure theory. Most of the results presented here are already known. The treatment of arbitrary real or complex Banach spaces as well as a number of simplified proofs are new.

Generalizations of the Banach-Stone theorem

The classical Banach-Stone theorem states that, for locally compact Hausdorff spaces M, N, the existence of an isometric isomorphism from C_oM onto C_oN implies that M and N are homeomorphic.
In part II of these notes we will consider the following vector-valued generalization of this theorem:

> Let X be a Banach space. Does it follow, for arbitrary locally compact Hausdorff spaces, that M and N are homeomorphic whenever $C_o(M,X)$ and $C_o(N,X)$ are isometrically isomorphic ?

(For generalizations of the Banach-Stone theorems in other directions we refer the reader to [5],[18],[20],[21],[22],[24],[28],[53],[65],[67],[74].)
Several authors have considered this problem (see chapter 8 for more details). We will show how it is dependent on the M-structure properties of X as to whether or not the vector-valued generalization of the Banach-Stone theorem is true. Our main results contain all

theorems known to the author which have been obtained without using M-structure methods as special cases.

The essential parts of our theory can be traced back to [10],[11], and [17]. The uniform treatment of both real and complex Banach spaces, the extension from compact to locally compact spaces, and the treatment of spaces with the local cns property (of which there are many more than the spaces considered in [17]) are new.

The contents of the individual chapters are as follows:

Part I: M-structure

Chapter 1: We study the basic properties of L-summands/M-summands and L-projections/M-projections (an L-projection (M-projection) is a projection E on a Banach space X such that $\|x\| = \|Ex\| + \|x-Ex\|$ ($\|x\| = \max\{\|Ex\|, \|x-Ex\|\}$) for every $x \in X$; L-summands (M-summands) are the ranges of L-projections (M-projections)).

Of particular importance for the following chapters will be the facts that L-projections and M-projections are mutually dual, that L-projections commute and that M-projections also commute, that the collection of L-projections (M-projections) on a Banach space forms a complete Boolean algebra (a Boolean algebra), and that a Banach space which is at least three-dimensional cannot have nontrivial L-projections and M-projections at the same time.

Chapter 2: In this chapter we treat M-ideals (an M-ideal is a closed subspace of a Banach space for which the annihilator is an L-summand in the dual space). We prove that CK-spaces have sufficiently many M-ideals (every closed ideal is an M-ideal and vice versa) and investigate the collection of all M-ideals of a Banach space. We also show that M-ideals can be characterized by certain intersection properties of balls.

Chapter 3: The aim of this chapter is the study of those operators T on a Banach space X for which every extreme functional is an eigenvector of the transposed operator T'. Such operators are called multipliers; we prove that they can be characterized by a certain boundedness condition and show how it is possible to get multipliers from bounded scalar-valued functions which are defined on the extreme functionals and continuous in the structure topology, a topology defined by means of the M-ideals of X. This result is a Banach space generalization of a well-known theorem for B^*-algebras, the Dauns-Hofmann theorem.

The centralizer of X, which will be denoted by Z(X), is the collection of those multipliers T for which a natural adjoint T^* exists (in the case of real scalars every multiplier is contained in Z(X)). We prove a number of characterization theorems for the operators in Z(X) and discuss some consequences of these results.

Chapter 4: We examine certain spaces of vector-valued functions (function modules) and the M-structure properties of such spaces. It is shown that every Banach space X can be regarded as a function module in such a way that the operators in the centralizer of X have a particularly simple form (the maximal function module representation). This representation has important consequences for the M-structure of X and will be fundamental for our generalizations of the Banach-Stone theorem.

Chapter 5: We investigate in more detail the M-structure properties of some special classes of Banach spaces (Banach spaces with finite dimensional centralizer, dual Banach spaces, B^*-algebras).

Chapter 6: This chapter contains bibliographical notes concerning chapters 1 - 5 as well as a number of supplements.

Part II: Generalizations of the Banach-Stone theorem

Chapter 7: We sketch some proofs of the classical Banach-Stone theorem and examine some simple consequences of this theorem.

Chapter 8: At the beginning of this chapter we give a precise definition of what is meant by a vector-valued generalization of the Banach-Stone theorem: We say that a Banach space X has the Banach-Stone property (the strong Banach-Stone property) if the existence of an isometric isomorphism from $C_o(M,X)$ onto $C_o(N,X)$ implies that M and N are homeomorphic (if every isometric isomorphism from $C_o(M,X)$ onto $C_o(N,X)$ has a certain simple form) whenever M and N are locally compact Hausdorff spaces ($C_o(M,X)$ denotes the space of X-valued continuous functions on M which vanish at infinity, provided with the supremum norm). We state some theorems of different authors concerning sufficient conditions for Banach spaces to have the (strong) Banach-Stone property which have been obtained without using M-structure methods and show how some of these theorems can easily be derived from the results in part I.

Chapter 9: In this chapter we examine Banach spaces for which the strong operator topology and the norm topology are equivalent on the centralizer (Banach spaces with a centralizer-norming system) and, more generally, Banach spaces for which the elements in the maximal function module representation do not behave too pathologically (Banach spaces with the local cns property). These results are important for the generalizations of the Banach-Stone theorem in chapter 11.

Chapter 10: We discuss systematically the M-structure properties of $C_o(M,X)$ (M a locally compact Hausdorff space, X a Banach space). The most important result is a theorem which states that a maximal function module representation of $C_o(M,X)$ can be obtained in a simple way if such a representation of X is known provided X has the local cns property.

Chapter 11: In this chapter we apply function module techniques to obtain explicit descriptions of isometric isomorphisms from $C_o(M,X)$ onto $C_o(N,Y)$ (M and N locally compact Hausdorff spaces, X and Y Banach spaces). Since the results of chapter 10 are essential for these investigations we only treat Banach spaces with the local cns property.

In the first part we derive sufficient conditions for the Banach-Stone property to hold. The second part is devoted to the study of Banach spaces whose centralizer is finite-dimensional. In this case we obtain necessary and sufficient conditions for the Banach-Stone property and the strong Banach-Stone property to hold. We show how the results which are mentioned in chapter 8 follow as corollaries from our theorems.

Chapter 12: In this final chapter we collect bibliographical notes, some supplements, and some open problems.

Thanks are due to R. Evans, H.-P. Butzmann, and P. Harmand for pointing out errors in the manuscript. I also wish to thank R. Evans and U. Schmidt-Bichler for the many helpful discussions during the preparation of these notes.

Finally, my gratitude goes to Mrs. Siewert who carefully inserted the symbols into the text.

C o n t e n t s

0. Preliminaries 1

A. Functional analysis 1

B. Topology 4

C. Boolean algebras 5

Part I: M-structure 7

1. L-projections and M-projections 8

A. Summands and projections 9

B. The structure of the collection of all L-summands and of all M-summands 19

C. L-summands and M-summands are $\mathbb{R}$-determined 22

D. The L-M-theorem 23

E. The Cunningham algebra 29

F. Summands in subspaces and quotients 31

2. M-ideals 33

A. The structure of the collection of M-ideals 34

B. A characterization of L-summands 41

C. A characterization of intersection properties of balls by properties of certain compact convex subsets in the dual space 44

D. A characterization of M-ideals by intersection properties 46

3. The centralizer 53

A. Multipliers and M-bounded operators 54

B. The centralizer 62

C. Characterization theorems 64

D. Applications of the characterization theorems 71

E. The space of primitive M-ideals 72

4. Function modules 75

A. General properties of function modules 77

B. Function module representations 90

C. Applications of the characterization theorems 104

5. M-structure of some classes of Banach spaces 108

A. Banach spaces for which the centralizer is one-dimensional 108

B. Banach spaces for which the centralizer is finite-dimensional 110

C. Dual Banach spaces 114
D. B^*-algebras 121

6. Remarks 122
Remarks concerning chapter 1 122
Supplement: L^p-summands and L^p-projections 122
Supplement: The Cunningham algebra 124
Remarks concerning chapter 2 125
Supplement: M-ideals and approximation theory 126
Remarks concerning chapter 3 128
Supplement: The centralizer of tensor products 129
Supplement: The bi-commutator of Z(X) 129
Remarks concerning chapter 4 130
Supplement: Function module techniques in approximation theory 131
Supplement: Square Banach spaces 132
Remarks concerning chapter 5 134

Part II: Generalizations of the Banach-Stone theorem 135

7. The classical Banach-Stone theorem 138

8. The Banach-Stone property and the strong Banach-Stone property 141

9. Centralizer-norming systems 152

10. M-structure of $C_o(M,X)$ 167
A. M-ideals and M-summands in $C_o(M,X)$ 167
B. The centralizer of $C_o(M,X)$ 169
C. Function module representations of $C_o(M,X)$ 171

11. Generalizations of the Banach-Stone theorem 178
A. The case of Banach spaces with the local cns property 179
B. The case of M-finite Banach spaces 190
C. Summary 198

12. Remarks 201
A. The Banach-Stone property for $\mathbb{C}$ 201
B. Square Banach spaces and centralizer-norming systems 203
C. Problems 207

Notation index 208
Subject index 210
References 212

0. Preliminaries

In this chapter we will collect together some well-known results from functional analysis, topology, and the theory of Boolean algebras which are essential for the following considerations.

A. Functional analysis

We will consider real or complex Banach spaces X, Y, ... The scalar field (which, as usual, is assumed to be fixed throughout every assertion) will be denoted by $\mathbb{K}$, where $\mathbb{K}$ is $\mathbb{R}$ or $\mathbb{C}$.
We will write $X \cong Y$ when the Banach spaces X, Y are isometrically isomorphic.

The usual notations of the classical Banach spaces will be adopted:

c_o = the space of all null sequences in $\mathbb{K}$

c = the space of all convergent sequences in $\mathbb{K}$

m = the space of all bounded sequences in $\mathbb{K}$

l^1 = the space of all absolutely convergent sequences in $\mathbb{K}$

l_n^∞ = the space $\mathbb{K}^n$, provided with the supremum norm

$C_o(M,X)$ = the space of X-valued continuous functions on M which vanish at infinity, provided with the supremum norm (M a locally compact Hausdorff space, X a Banach space)

$C_o(M,\mathbb{K})$ will be denoted by C_oM, and we will write $C(M,X)$ (and CM) instead of $C_o(M,X)$ (and C_oM) if M is compact.
If the scalar field is not arbitrary this will be expressed by writing $C_{\mathbb{R}}M$ and $C_{\mathbb{C}}M$ in the real and complex case, respectively.
C^bM denotes the space of bounded scalar-valued continuous functions on M.

An operator is a continuous linear map. The collection of all operators from X to Y (X and Y Banach spaces) will be denoted by [X,Y];

$[X,Y]_{iso}$ means the subset of isometrical isomorphisms in $[X,Y]$.
We will write X' and $B(X)$ instead of $[X,\mathbb{K}]$ and $[X,X]$.

There are a number of important locally convex topologies on $[X,Y]$.
We will only consider the

<u>strong operator topology</u> (which is defined to be the coarsest topology such that the mappings $T \mapsto Tx$ are continuous for every $x \in X$)

and the <u>weak operator topology</u> (the coarsest topology such that $T \mapsto p(Tx)$ is continuous for every $x \in X$, $p \in Y'$).

The case $Y = \mathbb{K}$, i.e. the weak*-topology on X', will be of particular importance in our investigations.

We assume the reader to be familiar with

1. the Hahn-Banach separation theorem
   (every nonvoid compact convex set in a locally convex Hausdorff space can be strictly separated from every disjoint nonvoid closed convex set)
2. the fact that the continuous linear functionals on $(X', \text{weak}^*\text{-topology})$ are just the evaluations at the points of X
3. the Alaoglu-Bourbaki theorem
   (the unit ball of X' is weak*-compact)
4. the Krein-Milman theorem
   (every compact convex set in a locally convex Hausdorff space is the closed convex hull of its extreme points)
5. the Krein-Šmulian theorem
   (a subspace of X' is weak*-closed iff its

intersection with the unit ball is weak*-closed)

(for proofs cf. e.g. [39], th. V.2.10, th. V.3.9, th. V.4.2, th. V.8.4, th. V.5.7).

Existence proofs for elements, balls, etc. in X can be derived from 1. and 2. (it might be necessary to apply 3. and 5. to show that the conditions of 1. are satisfied).

We will need the following version of the Hahn-Banach separation theorem which is an easy consequence of 1. and 2.:

0.1 Theorem: Let X be a Banach space and K_1, K_2 nonvoid convex subsets of $X' \times \mathbb{R}$ (X' provided with the weak*-topology).

Suppose that $K_1 \cap K_2 = \emptyset$ and that K_1 is closed and that K_2 is compact.

Then there are $x \in X$, $a, r \in \mathbb{R}$ such that

$$\operatorname{Re} p_1(x) + a a_1 < r < \operatorname{Re} p_2(x) + a a_2$$

for every $(p_1, a_1) \in K_1$, $(p_2, k_2) \in K_2$.

Note that $r > 0$ if $(0,0) \in K_1$ and that $a > 0$ if there is a $p \in X'$ and $a_1, a_2 \in \mathbb{R}$ such that $a_1 < a_2$, $(p, a_1) \in K_1$, $(p, a_2) \in K_2$.

Remark: We will make use of the elementary fact that the convex hull of a finite union of compact convex sets is compact. In our applications of th. 0.1 the sets K_1 and K_2 will sometimes be obtained in this way.

We will also need the following consequences of 1. and 2.:

- if $T: X \to Y$ is an operator and $T': Y' \to X'$ the transposed operator ($T'(p) := p \circ T$), then T' is weak*-continuous; conversely, every weak*-continuous operator $S: Y' \to X'$ has this form
- a subspace $\tilde{J}$ of X' is weak*-closed iff there is a subspace J of X such that $\tilde{J}$ is the annihilator of J in X', i.e. $\tilde{J} = J^{\pi} := \{p \mid p \in X', p|_J = 0\}$.

B. Topology

Topological spaces will be denoted by K, L, M, N, ... We will write $K \cong L$ when the topological spaces K and L are homeomorphic.

For $n \in \mathbb{N}$, nK means the disjoint union of n copies of K.

Let L be a locally compact Hausdorff space. αL and βL are the <u>one-point compactification</u> and the <u>Stone-Čech compactification</u> of L, respectively. We will only need the basic properties of these compactifications.

A compact space is called <u>extremally disconnected</u> if the closure of every open set is open. We note that every extremally disconnected space is totally disconnected and that a compact space K is extremally disconnected iff $C_{\mathbb{R}}K$ is a complete vector lattice.
An extremally disconnected space is said to be <u>hyperstonean</u> if the order continuous functionals $p \in (C_{\mathbb{R}}K)'$ separate the elements of $C_{\mathbb{R}}K$ ($p: C_{\mathbb{R}}K \to \mathbb{R}$ is <u>order continuous</u> if $p(\sup f_i) = \sup p(f_i)$ for every bounded increasing family $(f_i)_{i \in I}$ in $C_{\mathbb{R}}K$).

Let K be a compact Hausdorff space. We will need the following facts concerning CK and its subalgebras:

<u>0.2 Proposition</u>:

(i) Let A be a closed self-adjoint (i.e. $\overline{f} \in A$ for every $f \in A$) subalgebra of CK such that $\underline{1} \in A$.
Then $A = \{ f \circ \nu \mid f \in C(K/\sim) \}$, where $\nu : K \to K/\sim$ is the canonical map and $\sim$ is defined by "$k \sim l$ iff $f(k) = f(l)$ for every $f \in A$ "

(ii) Every closed subalgebra A of $C_{\mathbb{R}}K$ which contains $\underline{1}$ is a sublattice of $C_{\mathbb{R}}K$

(iii) Every homomorphism of B^*-algebras $p: CK \to \mathbb{K}$ is of the form δ_k ($\delta_k(f) := f(k)$) for a suitable $k \in K$

(iv) If $\omega : CK_1 \to CK_2$ is an isometric homomorphism of B^*-algebras

(K_1, K_2 compact Hausdorff spaces), then there is a continuous map t from K_2 onto K_1 such that $\omega(f) = f \circ t$ for every $f \in CK_1$

Proof:
(i) follows from the Stone-Weierstraß theorem, and (ii) is a corollary to (i). (iii) is a consequence of the fact that ker p is a maximal ideal in CK and thus of the form $\{ f \mid f(k) = 0 \}$ for a suitable $k \in K$. (iv) can easily be proved by using (iii). □

Partitions of unity will be of importance in our investigations in chapter 4 and chapter 10:

0.3 Proposition: Let L be a locally compact Hausdorff space and C a compact subset of L. If $U_1, \ldots, U_n$ are open subsets of L such that $C \subset U_1 \cup \ldots \cup U_n$, then there are continuous functions $h_1, \ldots, h_n$: $L \to [0,1]$ such that $\sum_{i=1}^{n} h_i|_C = 1$, $\underline{0} \le \sum_{i=1}^{n} h_i \le \underline{1}$, $h_i|_{L \setminus U_i} = 0$ $(i=1,\ldots,n)$

Proof: Choose a partition of unity $h_o, \ldots, h_n$ subordinate to the open cover $\alpha L \setminus C, U_1, \ldots, U_n$ of αL. Then $h_1, \ldots, h_n$ have the properties claimed. □

C. Boolean algebras

0.4 Definition: (see [55], p.5) A Boolean algebra $\mathfrak{B} = (B, \wedge, \vee, {}^\wedge, 0, 1)$ is a set B together with mappings $\wedge: B \times B \to B$, $\vee: B \times B \to B$, ${}^\wedge: B \to B$ and elements $0, 1 \in B$ such that the following conditions are satisfied:

$$p \wedge 1 = p \qquad p \vee 0 = p$$

$$p \wedge p^\wedge = 0 \qquad p \vee p^\wedge = 1$$

$$p \wedge q = q \wedge p \qquad p \vee q = q \vee p$$

$$p \wedge (q \vee r) = (p \wedge q) \vee (p \wedge r) \qquad p \vee (q \wedge r) = (p \vee q) \wedge (p \vee r) \qquad (\text{all } p,q,r \in B).$$

If $\mathfrak{B}$ is a Boolean algebra, then

" $p \le q$ iff $p \wedge q = p$ "

defines an order "$\le$" on B (the induced order). B is a lattice with respect to this ordering, and we say that $\mathfrak{B}$ is <u>complete</u> (<u>σ-complete</u>) if every subset of B (every countable subset of B) has a supremum and an infimum.

<u>0.5 Theorem</u>: For every Boolean algebra $\mathfrak{B} = (B,\wedge,\vee,^\wedge,0,1)$ there exists a totally disconnected compact Hausdorff space Ω (the <u>Stonean space</u> of $\mathfrak{B}$) and a map $\omega : B \to \{O \mid O \subset \Omega,\ O \text{ clopen}\}$ such that ω is one-to-one and the following conditions are satisfied:

$$\begin{aligned} \omega(p \wedge q) &= \omega(p) \cap \omega(q) \\ \omega(p \vee q) &= \omega(p) \cup \omega(q) \\ \omega(p^\wedge) &= \Omega \setminus \omega(p) \\ \omega(0) &= \emptyset \\ \omega(1) &= \Omega \qquad (\text{all } p,q \in B), \end{aligned}$$

i.e. every Boolean algebra can be regarded as the Boolean algebra of clopen subsets of a totally disconnected compact Hausdorff space.

Ω is uniquely determined by $\mathfrak{B}$ (up to homeomorphism), and $\mathfrak{B}$ is complete iff Ω is extremally disconnected.

<u>Proof</u>: [55], th. 6, th 10. □

PART I

M-STRUCTURE

1. L-projections and M-projections

The aim of this chapter is to investigate Banach spaces X which are direct sums $J \oplus J^{\perp}$ of subspaces J, $J^{\perp}$ such that, in addition, the norm of the sums $x+x^{\perp}$ is a simple function of $\|x\|$ and $\|x^{\perp}\|$ (all $x \in J$, $x^{\perp} \in J^{\perp}$). We will say that J is an L-summand if $\|x+x^{\perp}\| = \|x\| + \|x^{\perp}\|$ for $x \in J$, $x^{\perp} \in J^{\perp}$. J is called an M-summand if $\|x+x^{\perp}\| = \max\{\|x\|,\|x^{\perp}\|\}$ (all $x \in J$, $x^{\perp} \in J^{\perp}$).

In section A we consider some basic properties of such subspaces and the associated projections (L-projections and M-projections). L-projections and M-projections are mutually dual: the transpose of an L-projection is an M-projection and vice versa. This fact is of great technical importance and will be used to motivate the definition "M-ideal" in chapter 2. Every pair of L-projections or M-projections commute which implies that finite sums and intersections of L-summands (M-summands) are again L-summands (M-summands).

The structure of the collection of all L-projections and that of all M-projections is investigated further in section B. It turns out that the set of all L-projections is a complete Boolean algebra (the set of all M-projections is not complete in general). In section C we show that L-summands and M-summands of a complex Banach space can be determined in the underlying real space (as a corollary to this theorem we will get similar results for M-ideals and the centralizer in chapter 2 and chapter 3).

Section D contains a theorem which states that (apart from a single exception) a Banach space cannot have both nontrivial L-summands and M-summands at the same time. Using this theorem it is often easy to prove that a given space cannot have nontrivial L-summands (or M-summands, or M-ideals). In section E we prepare the investigations of the centralizer in chapter 3. We define the Cunningham algebra of a Banach space to be the Banach algebra generated by its

L-projections (a construction which is always of interest if one considers Boolean algebras of projections) and investigate some properties of this algebra.

Finally, in section F, we show how L-summands and M-summands can be determined in subspaces and quotients.

A. Summands and projections

1.1 Definition: Let X be a Banach space (the scalar field will be denoted by $\mathbb{K}$, where $\mathbb{K}$ is $\mathbb{R}$ or $\mathbb{C}$). A closed subspace J of X will be called an L-summand (M-summand) if there is a closed subspace $J^{\perp}$ of X such that X is the algebraic direct sum of J and $J^{\perp}$ and $\|x + x^{\perp}\| = \|x\| + \|x^{\perp}\|$ ($\|x + x^{\perp}\| = \max\{\|x\|, \|x^{\perp}\|\}$) for $x \in J$, $x^{\perp} \in J^{\perp}$.

The norm condition implies that the unit ball of X can be reconstructed in a simple way from the unit balls of J and $J^{\perp}$:

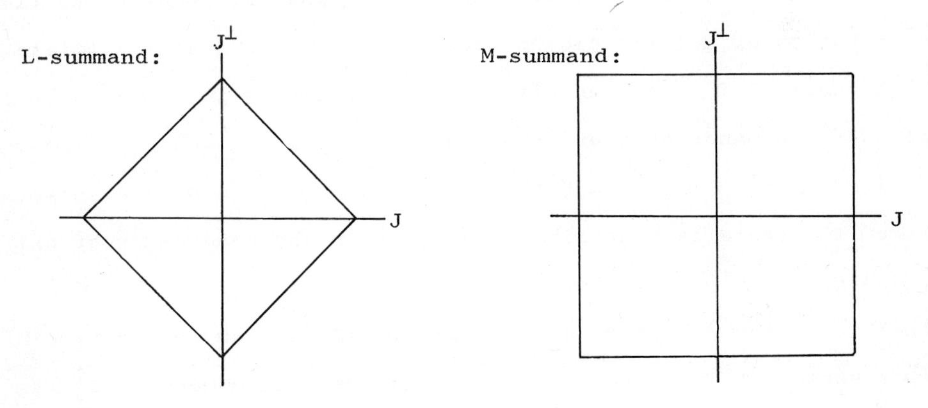

fig.1

Before we start to investigate the basic properties of L-summands and M-summands we give some examples:

1. It is clear that X and $\{0\}$ are always L-summands and M-summands.

These subspaces will be called the <u>trivial L- and M-summands</u>.

2. Let K be a locally compact Hausdorff space and C a clopen subset of K. Then, with $J_C := \{f \mid f \in C_oK, f|_C = 0\}$, it is easy to see that J_C is an M-summand in C_oK (take $(J_C)^{\perp} := J_{K \setminus C}$).
Conversely, it will be shown (cf. p. 13) that every M-summand of C_oK has this form. It should be noted that in the case of compact K this has alraedy been proved 1942 in a paper of Eilenberg ([41], th. 9.2).

3. For the classical sequence spaces c_o, c, and m we get the following as a corollary to the preceding result:

   a) the M-summands in c_o are the subspaces
   $\{(x_n) \mid (x_n) \in c_o, x_n = 0 \text{ for every } n \in D\}$, where D is a subset of $\mathbb{N}$ (since $c_o \cong C_o\mathbb{N}$)

   b) the M-summands in c are the subspaces
   $\{(x_n) \mid (x_n) \in c, x_n = 0 \text{ for every } n \in D\}$, where D is a finite or co-finite subset of $\mathbb{N}$
   (note that $c \cong C\alpha\mathbb{N}$, where $\alpha\mathbb{N}$ denotes the one-point compactification of $\mathbb{N}$; the clopen subsets of $\alpha\mathbb{N}$ are just the finite subsets of $\mathbb{N}$ together with their complements in $\alpha\mathbb{N}$)

   c) the M-summands of m are the subspaces
   $\{(x_n) \mid (x_n) \in m, x_n = 0 \text{ for every } n \in D\}$, where D is a subset of $\mathbb{N}$ (this is true since $m \cong C\beta\mathbb{N}$; an independent proof will be given on p. 18 below).

4. In $X = l^1$ the subspaces $J_D := \{(x_n) \mid (x_n) \in l^1, x_n = 0 \text{ for every } n \in D\}$ (D a subset of $\mathbb{N}$) are L-summands, and every L-summand of l^1 has this form (see p. 18). More generally, if (S, Σ, μ) is a measure space, then the annihilators of measurable subsets are L-summands in $L^1(S, \Sigma, \mu)$. The complete description of all L-summands is much more involved (cf. [16] , pp.56-59).

5. The following interesting example of an L-summand is due to Dixmier ([37], section 4):

Let H be a Hilbert space, K(H) the subspace of all compact operators in the space B(H) of all continuous linear operators on H. Dixmier proved that the annihilator of K(H) in the dual space of B(H) is an L-summand.

It is clear by the definition that, if J is an L-summand (M-summand), the space $J^\perp$ of def. 1.1 is also an L-summand (M-summand). We prove that $J^\perp$ is uniquely determined for J and that some of the conditions in definition 1.1 are in fact consequences of the norm condition.

1.2 Lemma:

(i) Let J be an L-summand (M-summand) in the Banach space X and suppose that the closed subspaces $J_1^\perp$, $J_2^\perp$ satisfy the conditions of def. 1.1. Then $J_1^\perp = J_2^\perp$.
We therefore are justified in calling $J^\perp$ the complementary L-summand (complementary M-summand) of J

(ii) If J is a subspace of the Banach space X such that there is a subspace $J^\perp$ for which $J + J^\perp = X$ and $\|x + x^\perp\| = \|x\| + \|x^\perp\|$ ($\|x + x^\perp\| = \max\{\|x\|, \|x^\perp\|\}$) for every $x \in J$, $x^\perp \in J^\perp$, then J is an L-summand (M-summand).

Proof:

(i) We prove that $J_1^\perp \subset J_2^\perp$ so that, by symmetry, $J_1^\perp = J_2^\perp$. For $y \in J_1^\perp$ we have $y = x + x^\perp$ with $x \in J$, $x^\perp \in J_2^\perp$. We consider the vector $ax + y$ for arbitrary $a \in \mathbb{R}$. Since J is an L-summand (M-summand) we get $|a|\|x\| + \|y\| = |a+1|\,\|x\| + \|x^\perp\|$ ($\max\{|a|\,\|x\|, \|y\|\} = \max\{|a+1|\,\|x\|, \|x^\perp\|\}$) for every $a \in \mathbb{R}$. This is possible only if $\|x\| = 0$, i.e. $y = x^\perp \in J_2^\perp$.

(ii) We have to show that $J \cap J^\perp = \{0\}$ and that J and $J^\perp$ are closed. Suppose that the norm condition $\|x + x^\perp\| = \|x\| + \|x^\perp\|$ (all $x \in J$, $x^\perp \in J^\perp$) is satisfied. For $x \in J \cap J^\perp$ we have $0 = \|0\| = \|x + (-x)\| = 2\|x\|$ so that $x = 0$. It follows that X is isometrically isomorphic to the product space $J \times J^\perp$, the product provided with the norm

$\|(x,x^\perp)\| := \|x\| + \|x^\perp\|$. Since X is complete, J and $J^\perp$ are also complete and therefore closed in X.

If $\|x + x^\perp\| = \max\{\|x\|,\|x^\perp\|\}$ (all $x \in J$, $x^\perp \in J^\perp$) the proof is similar.□

For technical reasons it will be convenient to translate the definitions of L-summands and M-summands into properties of projections.

<u>1.3 Definition</u>: Let X be a Banach space. A projection $E:X \to X$ (i.e. a continuous linear mapping E such that $E^2 = E$) is called an <u>L-projection</u> (<u>M-projection</u>) if $\|x\| = \|Ex\| + \|x - Ex\|$ ($\|x\| = \max\{\|Ex\|,\|x-Ex\|\}$) for every $x \in X$.

Note that L-projections and M-projections have norm less than or equal to one and that E is an L-projection (M-projection) iff Id-E is an L-projection (M-projection); "Id" denotes the identity operator from X to X.

It is not necessary to discuss examples since, as we shall see at once, projections and summands are in one-to-one correspondence. We only note that the operators 0 and Id are always L-projections and M-projections. They will be called the <u>trivial L- and M-projections</u>.

<u>1.4 Lemma</u>: Let $\mathbb{P}_L$ ($\mathbb{P}_M$) be the set of all L-projections (M-projections) on X (if several Banach spaces X are considered we will write $\mathbb{P}_L(X)$ and $\mathbb{P}_M(X)$ instead of $\mathbb{P}_L$ and $\mathbb{P}_M$).

(i) For $E \in \mathbb{P}_L$, range E and ker E are L-summands, and $E \leftrightarrow$ range E is a one-to-one correspondence between $\mathbb{P}_L$ and the set of all L-summands of X

(ii) For $E \in \mathbb{P}_M$, range E and ker E are M-summands, and $E \leftrightarrow$ range E is a one-to-one correspondence between $\mathbb{P}_M$ and the set of all M-summands of X

<u>Proof</u>:

(i) It follows at once from general properties of projections that

X is the algebraic direct sum of $J := \text{range } E$ and $J^{\perp} := \ker E$. For $x \in J$ and $x^{\perp} \in J^{\perp}$ we have

$$\begin{aligned} \|x + x^{\perp}\| &= \|E(x + x^{\perp})\| + \|(Id - E)(x + x^{\perp})\| \\ &= \|Ex\| + \|x^{\perp}\| \\ &= \|x\| + \|x^{\perp}\| \end{aligned}$$

so that J and $J^{\perp}$ are L-summands (note that J is closed since $J = \ker(Id - E)$).

Conversely, if J is an L-summand with $J^{\perp}$ as in def. 1.1, then it is clear that $x + x^{\perp} \mapsto x$ (all $x \in J$, $x^{\perp} \in J^{\perp}$) is an L-projection whose range is J.

(ii) can be proved analogously. □

As a first application of this translation technique we prove the result stated above in example 2:

<u>Example</u>: Let J be an M-summand in C_oK (K a locally compact Hausdorff space). Then there is a clopen subset C of K such that $J = J_C$ (cf. p.10)

<u>Proof</u>: Let E be the M-projection associated with J. For $f \in C_oK$ we have $\|2Ef - f\| = \max\{\|E(2Ef - f)\|, \|(Id - E)(2Ef - f)\|\} = \max\{\|Ef\|, \|(Id - E)(f)\|\} = \|f\|$, i.e. $F := 2E - Id$ is isometric. Since $F^2 = Id$ this map is in fact an isometric isomorphism so that, by the Banach-Stone theorem, there is a homeomorphism $t:K \to K$ and a continuous map $u:K \to \{a \mid a \in \mathbb{K}, |a| = 1\}$ such that $(Ff)(k) = u(k)f(t(k))$ for $f \in C_oK$, $k \in K$ (the Banach-Stone theorem will be the starting point of our investigations in part II of this volume; a proof is included in chapter 7).

It follows that α) $f(k) = u^2(k)f(t^2(k))$

β) $(Ef)(k) = 1/2[f(k) + u(k)f(t(k))]$

γ) $((Id - E)f)(k) = 1/2[f(k) - u(k)f(t(k))]$

for all $f \in C_oK$, $k \in K$ (since $F^2 = Id$, $E = 1/2(Id + F)$, $Id - E =$

$1/2(\mathrm{Id} - F)$).

β) and γ) imply that $t(k) = k$ for every $k \in K$: Suppose that there is a k_o such that $t(k_o) \neq k_o$. We choose a function $f \in C_o K$ such that $f(k)f(t(k)) = 0$ for every $k \in K$, $\|f\| = 1$. By β) and γ) we have $\|Ef\| \leq 1/2$, $\|(\mathrm{Id} - E)(f)\| \leq 1/2$ in contradiction to the norm condition for E.

α) implies that $u^2(k) = 1$, i.e. $u(k) = \pm 1$ for every k. Thus $Ef = \chi_D f$ (χ_D= the characteristic function of D) for every f, where D is the clopen set $\{ k \mid u(k) = 1 \}$.

It follows that $J = J_{K \setminus D}$. □

The following proposition is important for two reasons. First, it is often useful in order to get properties of M-summands from those of L-summands (or vice versa). Also, this proposition will be needed to motivate the definition "M-ideal" later.

<u>1.5 Proposition</u>: Let X be a Banach space and $E: X \to X$ a projection. Then (i) E is an L-projection iff E' is an M-projection

(ii) E is an M-projection iff E' is an L-projection

Since the annihilater of range E in X' is just the kernel of E' it follows from lemma 1.4 that the annihilator of an L-summand (M-summand) in X is an M-summand (L-summand) in X'.

<u>Proof</u>:

(i) Let E be an L-projection. It is clear that E' is a projection so that we only have to show that the norm condition holds:

$$\|p\| = \max\{\|p \circ E\|, \|p - p \circ E\|\} \qquad \text{for } p \in X'.$$

Because of $\|E\|, \|\mathrm{Id} - E\| \leq 1$ we have $\|p\| \geq \max\{\|p \circ E\|, \|p - p \circ E\|\}$.

For the proof of " $\leq$" we note that for $x \in X$, $\|x\| = 1$ and $p \in X'$ $|p(Ex)| \geq |p(x)| \|Ex\|$ or $|p((\mathrm{Id} - E)x)| \geq |p(x)| \|(\mathrm{Id} - E)x\|$ is valid (otherwise we would have $|p(x)| = |p(Ex + (\mathrm{Id} - E)x)|$

$$\leq |p(Ex)| + |p((\mathrm{Id} - E)x)|$$

$$< (\|Ex\| + \|(\mathrm{Id} - E)x\|)|p(x)|$$

$= |p(x)|$, a contradiction).

It follows easily that $\|p\| \le \max\{\|p\circ E\|, \|p - p\circ E\|\}$.

Now suppose that E' is an M-projection. For p, $q \in X'$ we then have

$$\|p\circ E + q\circ(Id - E)\| = \max\{\|p\circ E\|, \|q\circ(Id - E)\|\} .$$

Let $x \in X$ be arbitrarily given. We will prove that $\|x\| \ge \|Ex\| + \|(Id - E)x\|$ ("$\le$" is always valid). We choose $p,q \in X'$ such that $\|p\|,\|q\| = 1$, $p(Ex) = \|Ex\|$, $q((Id - E)x) = \|(Id - E)x\|$. It follows that

$$\begin{aligned} \|Ex\| + \|(Id - E)x\| &= p(Ex) + q((Id - E)x) \\ &\le \|p\circ E + q\circ(Id - E)\| \, \|x\| \\ &= (\max\{\|p\circ E\|, \|q\circ(Id - E)\|\})\|x\| \\ &\le \|x\| . \end{aligned}$$

(ii) Suppose that E is an M-projection. For $x, y \in X$ with $\|x\|,\|y\| \le 1$ it follows that $\|Ex + (Id - E)y\| = \max\{\|Ex\|, \|(Id - E)y\|\} \le 1$ so that, for $p \in X'$, $|(p\circ E)(x) + (p\circ(Id - E))(y)| = |p(Ex + (Id - E)(y))| \le \|p\|$. This implies that $\|p\circ E\| + \|p\circ(Id - E)\| \le \|p\|$, i.e. E' is an L-projection.

Conversely, if E' is an L-projection, then (by (i)) E'' is an M-projection on X'' . The claimed norm condition for E follows because we may regard X as a subspace of X'' and E as the restriction of E'' to X (similarly one can derive the second part of (i) from the first part of (ii)). □

<u>1.6 Corollary</u>:

(i) If X' (or X''', or $X^{(5)}$, ...) has no nontrivial L-summand, then X has no nontrivial M-summand

(ii) If X' (or X''', or $X^{(5)}$, ...) has no nontrivial M-summand, then X has no nontrivial L-summand □

A number of similar results will be proved later. They are of interest in connection with part II where it often is useful to know that a space has trivial M-structure (i.e. M-summands, M-ideals, centralizer).

We are now able to prove that L-projections commute and that M-projections commute. This fact will have a number of important consequences for the structure of the sets $\mathbb{P}_L$ and $\mathbb{P}_M$.

1.7 Proposition: Let X be a Banach space.

(i) If E and F are L-projections, then EF = FE, and EF and E+F-EF are also L-projections

(ii) If E and F are M-projections, then EF = FE, and EF and E+F-EF are also M-projections

Proof:

(i) Let E and F be L-projections. For $x \in X$ we have

$$\begin{aligned} \|Ex\| + \|x - Ex\| &= \|x\| \\ &= \|Fx\| + \|x - Fx\| \\ &= \|EFx\| + \|Fx - EFx\| + \|x - Fx\| \\ &\geq \|EFx\| + \|Fx - EFx\| + \|Ex - EFx\| \\ &\geq \|Ex\| + \|Fx - EFx\| \end{aligned}$$

so that $\|(Id - E)x\| \geq \|(Id - E)Fx\|$. If we replace E by Id - E in this inequality we get $\|Ex\| \geq \|EFx\|$ as well so that $EFx = FEx$ if $x \in$ range E or $x \in$ ker E. Since these subspaces span X it follows that EF = FE.

It remains to show that EF and E + F - EF are L-projections. Since EF = FE these mappings are projections. The norm conditions are proved as follows:

For $x \in X$ we have

α) $\|x\| = \|Ex\| + \|x - Ex\|$

β) $\|Ex\| = \|FEx\| + \|Ex - FEx\|$

γ) $\|x - EFx\| = \|Ex - EFx\| + \|x - Ex\|$

δ) $\|x - Ex\| = \|Fx - FEx\| + \|x - Ex - Fx + EFx\|$

ε) $\|Ex + Fx - EFx\| = \|Ex\| + \|Fx - EFx\|$.

A combination of α), β), and γ) gives $\|x\| = \|EFx\| + \|x - EFx\|$, and α), δ), and ε) yield $\|x\| = \|(E+F-EF)x\| + \|x - (E+F-EF)x\|$. Thus EF and E + F - EF are L-projections.

(ii) This follows at once from (i) and prop. 1.5. □

<u>1.8 Corollary</u>: Finite sums and intersections of L-summands (M-summands) are also L-summands (M-summands)

<u>Proof</u>: It suffices to consider the case of two summands. If E and F are the associated projections, then (range E) + (range F) = range (E + F - EF) and (range E) $\cap$ (range F) = range EF so that the assertions follow immediately from prop. 1.7. □

<u>Note</u>: Prop. 1.7 also implies that $(J_1 + J_2)^\perp = J_1^\perp \cap J_2^\perp$ and that $(J_1 \cap J_2)^\perp = J_1^\perp + J_2^\perp$ if J_1 and J_2 are L-summands or M-summands (this follows from (Id - E)(Id - F) = Id - (E+F-EF) and Id - EF = (Id-E) + (Id-F) - (Id-E)(Id-F) ; note that Id - E corresponds to $J^\perp$ if E corresponds to J).

We are now able to determine the L-summands and the M-summands in products of Banach spaces:

<u>L-summands in $\prod_{i\in I}^1 X_i$</u>: Let $(X_i)_{i\in I}$ be a family of Banach spaces and $\prod_{i\in I}^1 X_i$ their usual L^1-product (i.e. the set of those elements $(x_i)_{i\in I}$ in the product $\prod_{i\in I} X_i$ for which at most countably many x_i are different from zero and for which $\|(x_i)_{i\in I}\| := \sum_{i\in I} \|x_i\| < \infty$; we will denote all norms by the same symbol $\| \ \|$).

Then a closed subspace $J \subset \prod_{i\in I}^1 X_i$ is an L-summand iff $J = (\prod_{i\in I} J_i) \cap (\prod_{i\in I}^1 X_i)$, where J_i is an L-summand in X_i for every $i\in I$

<u>M-summands in $\prod_{i\in I}^\infty X_i$</u>: Let $(X_i)_{i\in I}$ be a family of Banach spaces and $\prod_{i\in I}^\infty X_i$ their usual Banach space product (i.e. the set of those elements $(x_i)_{i\in I}$ in the product $\prod_{i\in I} X_i$ for which $\|(x_i)_{i\in I}\| := \sup\{\|x_i\| \mid i\in I\} < \infty$).

Then a closed subspace $J \subset \prod\limits_{i\in I}^{\infty} X_i$ is an M-summand iff $J = (\prod\limits_{i\in I} J_i) \cap (\prod\limits_{i\in I}^{\infty} X_i)$, where J_i is an M-summand in X_i for every $i\in I$.

In particular, if $I = \mathbb{N}$ and every X_i is the scalar field $\mathbb{K}$, we obtain an explicit description of the L-summands in l^1 and the M-summands in m (cf. the examples on p.10).

Proof: We only give the proof for the case of L-summands in $\prod\limits_{i\in I}^{1} X_i$. The proof for M-summands is essentially the same.

In terms of projections we have to show that a projection $E: \prod\limits_{i\in I}^{1} X_i \to \prod\limits_{i\in I}^{1} X_i$ is an L-projection iff $E = \prod\limits_{i\in I} E_i$, where E_i is an L-projection on X_i for every $i\in I$ ($\prod\limits_{i\in I} E_i$ is the mapping $(x_i)_{i\in I} \mapsto (E_i x_i)_{i\in I}$).

It is clear that $\prod\limits_{i\in I} E_i \in \mathbb{P}_L(\prod\limits_{i\in I}^{1} X_i)$ if every $E_i \in \mathbb{P}_L(X_i)$.

Conversely, let E be an L-projection on $\prod\limits_{i\in I}^{1} X_i$. For $i\in I$, we denote by π_i and j_i the natural i'th projection (from $\prod\limits_{i\in I}^{1} X_i$ onto X_i) and the natural i'th injection (from X_i into $\prod\limits_{i\in I}^{1} X_i$), respectively.

It is clear from the definition of the norm in $\prod\limits_{i\in I}^{1} X_i$ that $j_i\pi_i$ is an L-projection so that $E(j_i\pi_i) = (j_i\pi_i)E$. Since $j_i\pi_i j_i = j_i$ it follows that $Ej_i = j_i\pi_i E j_i$.

For $i\in I$ we define $E_i := \pi_i E j_i$. We will prove that these mappings are L-projections and that $E = \prod\limits_{i\in I} E_i$.

Every E_i is idempotent since

$$\begin{aligned} E_i^2 &= \pi_i E j_i \pi_i E j_i \\ &= \pi_i E^2 j_i \\ &= \pi_i E j_i \\ &= E_i \,. \end{aligned}$$

For $i\in I$ and $x_i\in X_i$ we have

$$\begin{aligned} \|x_i\| &= \|j_i x_i\| \\ &= \|E j_i x_i\| + \|j_i x_i - E j_i x_i\| \\ &= \|j_i \pi_i E j_i x_i\| + \|j_i x_i - j_i \pi_i E j_i x_i\| \\ &= \|j_i E_i x_i\| + \|j_i (x_i - E_i x_i)\| \\ &= \|E_i x_i\| + \|x_i - E_i x_i\| . \end{aligned}$$

Finally we have to show that $E = \prod_{i\in I} E_i$. It suffices to prove that $\pi_{i_o} E = \pi_{i_o} \prod_{i\in I} E_i \; (= E_{i_o} \pi_{i_o})$ for every $i_o \in I$. But this follows easily from $\pi_{i_o} j_{i_o} \pi_{i_o} = \pi_{i_o}$:

$$\begin{aligned} E_{i_o} \pi_{i_o} &= \pi_{i_o} E(j_{i_o} \pi_{i_o}) \\ &= \pi_{i_o} (j_{i_o} \pi_{i_o}) E \\ &= \pi_{i_o} E \end{aligned}$$

□

B. The structure of the collection of all L-summands and of all M-summands

The results of section A imply that $\mathbb{P}_L$ and $\mathbb{P}_M$ are Boolean algebras in a natural way:

1.9 Lemma: (cf. chapter 0 for definitions) If E and F are L-projections (or M-projections) on the Banach space X, we define:

$$E \wedge F := EF$$

$$E \vee F := E + F - EF$$

$$E^\wedge := Id - E \quad .$$

With these definitions, $\mathbb{P}_L$ and $\mathbb{P}_M$ are Boolean algebras with maximal element Id and minimal element 0.

Since projections and summands are in one-to-one correspondence, the set of all L-summands and the set of all M-summands are also Boolean algebras. For L-summands or M-summands J, J_1, J_2 we have

$$J_1 \wedge J_2 = J_1 \cap J_2$$

$$J_1 \vee J_2 = J_1 + J_2$$

$$J^\wedge = J^\perp \quad ,$$

and the induced order ($J_1 \le J_2$ iff $J_1 \wedge J_2 = J_1$) is just the inclusion order.

Proof: The proof is routine. □

As Boolean algebras $\mathbb{P}_L$ and $\mathbb{P}_M$ are lattices with respect to the induced order. We will see that $\mathbb{P}_L$ is in fact a complete Boolean algebra and that $\mathbb{P}_M$ is not complete in general (note that this is our first result where L-projections and M-projections do not have similar properties).

We recall that a family $\underline{E}$ in an ordered space is called <u>increasing</u> if for $E_1, E_2 \in \underline{E}$ there is an $E_3 \in \underline{E}$ such that $E_1, E_2 \leq E_3$. We say that an increasing family of L-projections is pointwise convergent to a map $T: X \to X$, if $\lim\limits_{E \in \underline{E}} Ex = Tx$ for every $x \in X$ (i.e. for every $\varepsilon > 0$ there is an $E_o \in \underline{E}$ such that $\|Ex - Tx\| \leq \varepsilon$ for every $E \in \underline{E}$, $E \geq E_o$).

<u>1.10 Theorem</u>: Let X be a Banach space. Then

(i) Every increasing family $\underline{E}$ of L-projections is pointwise convergent to an L-projection E_o.

E_o is the supremum of the family $\underline{E}$ in the Boolean algebra $\mathbb{P}_L$

(ii) $\mathbb{P}_L$ is a complete Boolean algebra

(iii) $\mathbb{P}_M$ is not complete in general

<u>Proof</u>:

(i) Let $\underline{E}$ be an increasing family of L-projections and $x \in X$. Since $E \leq F$ $(E, F \in \mathbb{P}_L)$ implies that $\|Fx\| = \|Ex\| + \|Fx - Ex\|$, the family $(\|Ex\|)_{E \in \underline{E}}$ is also increasing in $\mathbb{R}$ and bounded from above. We choose $E_1, E_2, \ldots \in \underline{E}$ such that $E_1 \leq E_2 \leq \ldots$ and $\lim\limits_{n \to \infty} \|E_n x\| = \sup\limits_{E \in \underline{E}} \|Ex\|$.

We have $\|E_n x - E_m x\| = \|E_n x\| - \|E_m x\|$ for $n, m \in \mathbb{N}$, $n \geq m$, so that $(E_n x)_{n \in \mathbb{N}}$ is a Cauchy sequence.

We define $E_o: X \to X$ by $E_o x := \lim\limits_{n \to \infty} E_n x$ (note that the sequence $(E_n)_{n \in \mathbb{N}}$ depends on x). It remains to prove that

$$\alpha)\ E_o x = \lim_{E \in \underline{E}} Ex \quad \text{for } x \in X$$

$$\beta)\ E_o \text{ is an L-projection}$$

$$\gamma)\ E_o = \sup_{E \in \underline{E}} E .$$

α) For $\varepsilon > 0$ we choose an E_n such that $\|E_o x - E_n x\| \leq \varepsilon$ and $\|E_n x\| \geq \sup\limits_{E \in \underline{E}} \|Ex\| - \varepsilon$. But then, for $E \in \underline{E}$ such that $E \geq E_n$ it follows that $\|Ex - E_o x\| \leq \|Ex - E_n x\| + \|E_n x - E_o x\| \leq \|Ex\| - \|E_n x\| + \varepsilon \leq 2\varepsilon$.

β) By α), E_o is a linear map which (as the pointwise limit of L-projections) satisfies the norm condition $\|x\| = \|E_o x\| + \|x - E_o x\|$ for every x. We thus merely have to prove that $E_o^2 = E_o$.

To this end, let $\varepsilon > 0$ and $x \in X$. Choose $\tilde{E} \in \underline{E}$ such that $\|EE_o x - E_o^2 x\| \leq \varepsilon$ for every $E \in \underline{E}$, $E \geq \tilde{E}$. For sufficiently large E (wlog $E \geq \tilde{E}$) we have further that $\|Ex - E_o x\| \leq \varepsilon$ and therefore that $\|Ex - EE_o x\| = \|E(Ex - E_o x)\| \leq \varepsilon$. Hence $\|E_o x - E_o^2 x\| \leq 3\varepsilon$ so that $E_o = E_o^2$.

γ) We have to show that $EE_o = E$ for every $E \in \underline{E}$ and that $E_o F = E_o$ provided that $EF = E$ for every $E \in \underline{E}$.

But this follows easily from α).

(ii) It is wellknown that a Boolean algebra is complete iff it contains suprema of increasing families (if $\underline{E}$ is any family in $\mathbb{P}_L$, then $\sup \underline{E} = \sup \underline{E}^{\vee}$, where $\underline{E}^{\vee}$ denotes the increasing family $\{E_1 \vee E_2 \vee \ldots \vee E_n \mid n \in \mathbb{N},\ E_1, \ldots, E_n \in \underline{E}\}$; note further that $\inf \underline{E} = \mathrm{Id} - (\sup_{E \in \underline{E}} (\mathrm{Id} - E))$).

(iii) For every locally compact Hausdorff space K the Boolean algebra of clopen subsets of K and $\mathbb{P}_M(C_o K)$ are isomorphic as Boolean algebras ($C \mapsto J_{K \setminus C}$ is such an isomorphism; cf. example 2 on p.10). It is easy to see that the Boolean algebra of clopen subsets is not complete in general (for example, if $K = \alpha\mathbb{N}$, the sequence $C_n := \{2,4,\ldots,2n\}$, $n=1,2,\ldots$, has no supremum). □

<u>Note</u>: We will see later (theorem 5.7(ii)) that $\mathbb{P}_M$ is complete for every dual Banach space.

As a corollary we obtain

<u>1.11 Theorem</u>: Let $\underline{J}$ be a family of L-summands in the Banach space X. Then (i) $\overline{\mathrm{lin}}\,(\bigcup_{J \in \underline{J}} J)$ is an L-summand, and $(\overline{\mathrm{lin}} \bigcup_{J \in \underline{J}} J)^{\perp} = \bigcap_{J \in \underline{J}} J^{\perp}$

(ii) $\bigcap_{J \in \underline{J}} J$ is an L-summand, and $(\bigcap_{J \in \underline{J}} J)^{\perp} = \overline{\mathrm{lin}}(\bigcup_{J \in \underline{J}} J^{\perp})$

<u>Proof</u>: Let $\underline{E}$ be the family of L-projections which corresponds to $\underline{J}$:

(i) Without loss of generality we may assume that the family $\underline{J}$ is increasing so that $J_o := \overline{\text{lin}} \bigcup_{J \in \underline{J}} J = (\bigcup_{J \in \underline{J}} J)^-$. Then $\underline{E}$ is also increasing, and $E_o := \sup_{E \in \underline{E}} E$ is the pointwise limit of the $E \in \underline{E}$.

We claim that J_o = range E_o. Since $E \le E_o$ for every $E \in \underline{E}$ it follows that range $E \subset$ range E_o so that $J_o = (\bigcup_{E \in \underline{E}} \text{range } E)^- \subset \text{range } E_o$.

Conversely, for $x \in$ range E_o, we have $x = \lim_{E \in \underline{E}} Ex$ and consequently $x \in J_o$.

It remains to show that $\ker E_o = \bigcap_{J \in \underline{J}} J^{\perp} (= \bigcap_{E \in \underline{E}} \ker E)$.

The inclusion "$\supset$" is an immediate consequence of 1.10(i). Conversely, let $x \in X$ be given such that $x \notin \bigcap_{E \in \underline{E}} \ker E$, i.e. $\tilde{E}x \neq 0$ for some $\tilde{E} \in \underline{E}$. Since $\|Ex\| \ge \|\tilde{E}x\|$ for $E \ge \tilde{E}$ and $E_o x = \lim_{E \in \underline{E}} Ex$ it follows that $\|E_o x\| \ge \|\tilde{E}x\| > 0$ so that $x \notin \ker E_o$.

(ii) We only have to apply (i) to the family $(J^{\perp})_{J \in \underline{J}}$. □

Note: The intersection and the closed linear span of M-summands are not M-summands in general (cf. the counterexamples in 1.10(iii)).

C. L-summands and M-summands are $\mathbb{R}$-determined

We will say that a class of subspaces or operators is $\mathbb{R}$-determined if, for complex spaces, the relevant subspaces or operators can be determined in the underlying real space.

1.12 Theorem: Let X be a complex Banach space.

(i) If J is an L-summand in $X_{\mathbb{R}}$ (= the underlying real space), then J is an L-summand in X

(ii) X and $X_{\mathbb{R}}$ have the same L-summands, M-summands, L-projections, and M-projections

Proof:

(i) It suffices to prove that $ix_o \in J$ if $x_o \in J$. Define $y_o := x_o + iEix_o$, where E is the L-projection (on $X_{\mathbb{R}}$) which corresponds to J. An easy computation shows that $x \mapsto -iEix$ defines an L-projection $\tilde{E}$ (on $X_{\mathbb{R}}$) so that $E\tilde{E} = \tilde{E}E$. This yields

$$\begin{aligned} Ey_o &= E(E - \tilde{E})x_o \\ &= (E - \tilde{E})Ex_o \\ &= (E - \tilde{E})x_o \\ &= y_o . \end{aligned}$$

We have also $E(iy_o) = 0$ by definition so that

$$\begin{aligned} |1+i|\,\|y_o\| &= \|y_o + iy_o\| \\ &= \|E(y_o+iy_o)\| + \|(Id-E)(y_o+iy_o)\| \\ &= \|y_o\| + \|iy_o\| \\ &= 2\|y_o\| . \end{aligned}$$

This is possible only if $\|y_o\| = 0$ so that necessarily $E(ix_o) = ix_o$, i.e. $ix_o \in J$.

(ii) It is clear that L-summands (M-summands, L-projections, M-projections) in X are also L-summands (M-summands, L-projections, M-projections) in $X_{\mathbb{R}}$. The converse follows from (i), 1.4, and 1.5 (note that there is a natural isometric isomorphism from $(X')_{\mathbb{R}}$ onto $(X_{\mathbb{R}})'$, $p \mapsto \operatorname{Re} p$). □

D. The L-M-theorem

When we considered the examples above we investigated in each case only one type of summands or projections. This is justified by the main theorem of this section:

We will prove that, up to a single exception, a Banach space cannot have nontrivial L-summands and M-summands at the same time. The exceptional case is the two-dimensional space $\mathbb{R}^2$, provided with the supremum norm (this space will be denoted by l_2^{∞}) which has two obvious nontrivial L-summands and two obvious nontrivial M-summands:

the space l_2^∞ :

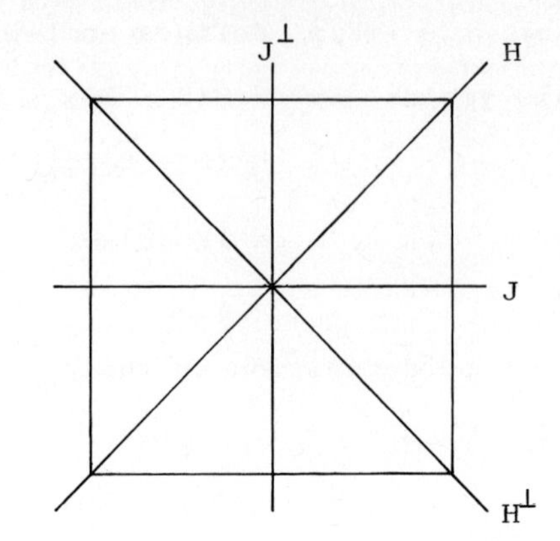

$J, J^\perp$: M-summands

$H, H^\perp$: L-summands

fig. 2

1.13 Theorem: Let X be a Banach space.

(i) If $\mathbb{K} = \mathbb{C}$, then all L-summands of X or all M-summands of X are trivial

(ii) If $\mathbb{K} = \mathbb{R}$ and X is not isometrically isomorphic to the real Banach space l_2^∞ , then all L-summands or all M-summands of X are trivial

Proof: Since l_2^∞ is not isometrically isomorphic to the underlying real space of any complex space it is sufficient to prove (ii).

We assume that X has a nontrivial L-summand H and a nontrivial M-summand J. We will prove that J, $J^\perp$, H, and $H^\perp$ are one-dimensional so that X and l_2^∞ are isometrically isomorphic.

The proof is rather technical (unfortunately, the author knows no simpler proof), the idea, however, is simple:

It will be shown in several steps that the J, $J^\perp$, H, $H^\perp$ behave as indicated in fig. 2, namely

α) $J \cap H = J \cap H^\perp = J^\perp \cap H = J^\perp \cap H^\perp = \{0\}$

β) let $y \in H$ (or, analogously, $y \in H^\perp$) be decomposed as $y = z + z^\perp$ where $z \in J, z^\perp \in J^\perp$; then $\|y\| = \|z\| = \|z^\perp\|$

γ) if $z \in J$ (or $z \in J^\perp$) is decomposed as $z = y + y^\perp$

with $y \in H$, $y^\perp \in H^\perp$, then $\|y\| = \|y^\perp\| = \|z\|/2$

δ) for $y \in H$ and z, $z^\perp$ as in β) we have $z - z^\perp \in H^\perp$

ε) $\dim J = \dim J^\perp = \dim H = \dim H^\perp = 1$.

α) We only have to show that $J \cap H = \{0\}$; the proofs for $J \cap H^\perp$, $J^\perp \cap H$, and $J^\perp \cap H^\perp$ are analogous.

Suppose that there is a vector $z_o \in J \cap H$, $\|z_o\| = 1$. We will prove that $J^\perp \subset H$ in this case.

For $z^\perp \in J^\perp$, $\|z^\perp\| = 1$, we write $z^\perp = y + y^\perp$, where $y \in H$, $y^\perp \in H^\perp$, $\|y\| + \|y^\perp\| = 1$. We decompose $z_o \pm z^\perp$ $(= (z_o \pm y) \pm y^\perp)$ with respect to $H, H^\perp$: $1 = \max\{\|z_o\|, \|\pm z^\perp\|\} = \|z_o \pm z^\perp\| = \|z_o \pm y\| + \|y^\perp\|$.

It follows that $\|z_o \pm y\| = 1 - \|y^\perp\| = \|y\|$ so that

$$
\begin{aligned}
1 &= \|z_o\| \\
&= \|1/2(z_o+y) + 1/2(z_o-y)\| \\
&\leq \|y\|
\end{aligned}
$$

and consequently $\|y\| = 1$, i.e. $y^\perp = 0$ and $z^\perp = y \in H$.

In particular we have $J^\perp \cap H \neq \{0\}$ (since $J^\perp \subset H$), and a similar proof yields $J \subset H$ (replace $J \cap H$ by $J^\perp \cap H$).

But then $X = J + J^\perp = H$ in contrast to the assumption that H is a nontrivial L-summand.

β) Since J is an M-summand, we have $\|y\| = \max\{\|z\|, \|z^\perp\|\}$. Without loss of generality we may assume that $\|y\| = \|z\| = 1$, and it remains to show that $\|z^\perp\| = 1$, too.

Suppose that $\|z^\perp\| < 1$. We decompose z and $z^\perp$ with respect to $H, H^\perp$: $z = y_1 + y_1^\perp$, $z^\perp = y_2 + y_2^\perp$ $(y_1, y_2 \in H$, $y_1^\perp, y_2^\perp \in H^\perp)$. Since $y = (y_1+y_2) + (y_1^\perp + y_2^\perp)$ and $H \cap H^\perp = \{0\}$ it follows that $y_1^\perp = -y_2^\perp$, i.e. $z = y_1 + y_1^\perp$ and $z^\perp = y_2 - y_1^\perp$.

We now consider $z(a) := z + az^\perp$ for $a \in \mathbb{R}$. We have $\|z(a)\| = \max\{\|z\|, |a|\,\|z^\perp\|\} = 1$ for a in a suitable nondegenerate interval $[1-\varepsilon, 1+\varepsilon]$ so that $1 = \|z(a)\| = \|(y_1+ay_2)+(1-a)y_1^\perp\| = \|y_1+ay_2\| + |1-a|\,\|y_1^\perp\|$ for these a. But then

$$
\begin{aligned}
\|y\| &= \|y_1 + y_2\| \\
&= \|1/2(y_1+(1+\varepsilon)y_2) + 1/2(y_1+(1-\varepsilon)y_2)\| \\
&\leq 1/2\|y_1+(1+\varepsilon)y_2\| + 1/2\|y_1+(1-\varepsilon)y_2\| \\
&= 1 - \varepsilon\|y_1^{\perp}\|
\end{aligned} ,
$$

and the last expression is strictly less than one (note that, by α), any nonvanishing vector in H or J has nonvanishing components in J, $J^{\perp}$ or H, $H^{\perp}$, respectively, so that in particular $y_1^{\perp} \neq 0$). This contradiction proves β).

γ) By β) we may decompose y and $y^{\perp}$ as

$$
\begin{aligned}
y &= z_1 + z_1^{\perp} \\
y^{\perp} &= z_2 + z_2^{\perp} ,
\end{aligned}
$$

where $z_1, z_2 \in J$, $z_1^{\perp}, z_2^{\perp} \in J^{\perp}$, $\|z_1\| = \|z_1^{\perp}\| = \|y\|, \|z_2\| = \|z_2^{\perp}\| = \|y^{\perp}\|$. Since $z = (z_1+z_2) + (z_1^{\perp} + z_2^{\perp})$ and $J \cap J^{\perp} = \{0\}$ it follows that $z_1^{\perp} + z_2^{\perp} = 0$ so that $\|y\| = \|z_1^{\perp}\| = \|z_2^{\perp}\| = \|y^{\perp}\|$.

This proves that $\|y\| = \|y^{\perp}\| = 1/2\|z\|$.

δ) Let $y \in H$ be given. With the notation of the proof of β) we have

$$
\begin{aligned}
y &= z + z^{\perp} \\
z &= y_1 + y_1^{\perp} \\
z^{\perp} &= y_2 - y_1^{\perp}
\end{aligned}
$$

where, by γ), $\|y_1\| = \|y_1^{\perp}\| = \|y_2\| = 1/2\|y\|$. But then

$$
\begin{aligned}
\|y\| &= \|z + z^{\perp}\| \\
&= \max\{\|z\|, \|z^{\perp}\|\} \\
&= \|z - z^{\perp}\| \\
&= \|y_1 - y_2 + 2y_1^{\perp}\| \\
&= \|y_1 - y_2\| + 2\|y_1^{\perp}\| \\
&= \|y_1 - y_2\| + \|y\|
\end{aligned}
$$

so that $y_1 - y_2 = 0$ and consequently $z - z^{\perp} = 2y_1^{\perp} \in H^{\perp}$.

ε) It is sufficient to prove that $\dim H = 1$ (analogously one shows that $\dim H^{\perp} = 1$ so that $\dim X = 2$ and $\dim J = \dim J^{\perp} = 1$).

Suppose that $\dim H \geq 2$. We choose $y_1, y_2 \in H$ such that $\|y_1\| = \|y_2\| = 1$,

$y_1 \pm y_2 \neq 0$. For $i = 1,2$ we write $y_i = z_i + z_i^{\perp}$ with $z_i \in J$, $z_i^{\perp} \in J^{\perp}$. By β) and δ) we know that $\|z_i\| = \|z_i^{\perp}\| = 1$ and that $y_i^{\perp} :=$ $z_i - z_i^{\perp} \in H^{\perp}$.

We claim that $\{z_1, z_2\}$ and $\{z_1^{\perp}, z_2^{\perp}\}$ are linearly independent. Since these vectors are normalized this is equivalent to $z_1 \pm z_2 \neq 0$, $z_1^{\perp} \pm z_2^{\perp} \neq 0$, and this follows from α) and $y_1 \pm y_2 \neq 0$.

The vectors $z_1, z_2, z_1^{\perp}, z_2^{\perp}$ generate a fourdimensional subspace of X for which $\operatorname{lin}\{z_1, z_2\}$ is a nontrivial M-summand (with complementary M-summand $\operatorname{lin}\{z_1^{\perp}, z_2^{\perp}\}$) and for which $\operatorname{lin}\{y_1, y_2\}$ is a nontrivial L-summand (with complementary L-summand $\operatorname{lin}\{y_1^{\perp}, y_2^{\perp}\}$). We will prove that this is impossible.

First note that

$$\begin{aligned} \|z_1 \pm z_2\| &= \|1/2(y_1 + y_1^{\perp}) \pm 1/2(y_2 + y_2^{\perp})\| \\ &= \|1/2(y_1 - y_1^{\perp}) \pm 1/2(y_2 - y_2^{\perp})\| \\ &= \|z_1^{\perp} \pm z_2^{\perp}\| \end{aligned}$$

(since $\|y + y^{\perp}\| = \|y - y^{\perp}\|$ for $y \in H$, $y^{\perp} \in H^{\perp}$). It follows that

$$\begin{aligned} 2 &= \|y_1\| + \|y_2^{\perp}\| \\ &= \|y_1 + y_2^{\perp}\| \\ &= \|z_1 + z_1^{\perp} + z_2 - z_2^{\perp}\| \\ &= \max\{\|z_1 + z_2\|, \|z_1^{\perp} - z_2^{\perp}\|\} \\ &< \|z_1 + z_2\| + \|z_1^{\perp} - z_2^{\perp}\| \\ &= \max\{\|z_1 + z_2\|, \|-(z_1^{\perp} + z_2^{\perp})\|\} + \max\{\|z_1 - z_2\|, \|z_1^{\perp} - z_2^{\perp}\|\} \\ &\qquad \text{(this follows from } \|z_1 \pm z_2\| = \|z_1^{\perp} \pm z_2^{\perp}\|) \\ &= \|z_1 + z_2 - z_1^{\perp} - z_2^{\perp}\| + \|z_1 - z_2 + z_1^{\perp} - z_2^{\perp}\| \\ &= \|y_1^{\perp} + y_2^{\perp}\| + \|y_1 - y_2\| \\ &= \|(y_1 - y_2) + (y_1^{\perp} + y_2^{\perp})\| \\ &= \|2z_1 - 2z_2^{\perp}\| \\ &= \max\{\|2z_1\|, \|2z_2^{\perp}\|\} \\ &= 2 \end{aligned}$$

This contradiction shows that H is one-dimensional, and the proof is complete. □

1.14 Corollary: If X is a Banach space which is not isometrically isomorphic to the real Banach space l_2^∞, then the following hold:

(i) If X' (or X''', or $X^{(5)}, \ldots$) contains a nontrivial L-summand (M-summand), then X contains no nontrivial L-summand (M-summand)

(ii) If X (or X'', or $X^{(4)}, \ldots$) contains a nontrivial L-summand (M-summand), then X contains no nontrivial M-summand (L-summand)

Proof: This follows at once from th. 1.13 and cor. 1.6 . □

We give some applications:

1. c_o, c, and m have no nontrivial L-summands

2. Let X be a Lindenstrauß space which is not isometrically isomorphic to the real Banach space l_2^∞. Then X has no nontrivial L-summands.

   (A Lindenstrauß space is a Banach space for which the dual space is isometrically isomorphic to a space $L^1(S,\Sigma,\mu)$. Note that $L^1(S,\Sigma,\mu)$ always has nontrivial L-summands if this space is not one-dimensional.)

   In particular, C_oK has no nontrivial L-summands if K contains more than two points(K a locally compact Hausdorff space).

3. The only real (resp. complex) Banach spaces X which are at the same time a Lindenstrauß space and a space $L^1(S,\Sigma,\mu)$ are $\mathbb{R}$ and l_2^∞ (resp. $\mathbb{C}$)

4. Let H be a Hilbert space. Then B(H) and K(H) have no nontrivial L-summands

   (this follows from example 5, p. 10, and the fact that $(K(H))'' \cong B(H)$; see [37])

5. A product $\prod^1_{i\in I} X_i$ ($\prod^\infty_{i\in I} X_i$) has no nontrivial M-summands (L-summands) provided that the product is not isometrically isomorphic to the real space l_2^∞ and there are at least two indices $i\in I$ such that $X_i \neq \{0\}$.

E. The Cunningham algebra

The generated Banach algebra plays an important role in the theory of Boolean algebras of projections (for general results in this direction see [40], chapter XVII). For the case of L-projections this algebra has been studied extensively by Alfsen and Effros. We follow their terminology in calling this algebra the "Cunningham algebra".

1.15 Definition: Let X be a Banach space. Then the Cunningham algebra, C(X), is the Banach algebra generated by $\mathbb{P}_L$ in B(X). It is clear that C(X) is the closure of the linear span of $\mathbb{P}_L$.
(We hope that the reader is not confused by our using the same symbol "C" to denote the Cunningham algebra C(X) for Banach spaces X and the space of continuous functions CK for compact Hausdorff spaces K; there will be no assertions where confusion is possible.)

Example: Since the L-projections on l^1 are in one-to-one correspondence with the idempotents of the Banach algebra m (see p. 10) it follows easily that the mapping

$$(y_n) \to ((x_n) \to (y_n x_n))$$

is an isometric isomorphism from m onto $C(l^1)$ so that $C(l^1) \cong m$.

Since $\mathbb{P}_L$ is a complete Boolean algebra there is an extremally disconnected compact Hausdorff space Ω_X and an isomorphism of Boolean algebras, ω_X, from the set of clopen subsets of Ω_X onto $\mathbb{P}_L$. If we identify the clopen subsets of Ω_X with their associated characteristic functions in $C(\Omega_X)$ we may regard ω_X as a map from a subset of $C(\Omega_X)$ onto $\mathbb{P}_L$. We then have

1.16 Proposition: The mapping ω_X has a unique extension to an isometric algebra isomorphism from $C(\Omega_X)$ onto C(X).
This extension will also be denoted by ω_X.

Proof:

Uniqueness is clear since $C(\Omega_X)$ is the closed linear span of the characteristic functions contained in it.

We define ω_X on the linear span of the characteristic functions by

$$\omega_X\Big(\sum_{i=1}^{n} a_i\, \chi_{O_i}\Big) := \sum_{i=1}^{n} a_i\, \omega_X(O_i)$$

($O_1,\dots,O_n$ clopen subsets of Ω_X, $a_1,\dots,a_n \in \mathbb{K}$).

We first note that ω_X is well-defined. This depends on the fact that $\omega_X(O_1 \cup O_2) = \omega_X(O_1) + \omega_X(O_2)$ for disjoint clopen subsets of Ω_X. We omit the elementary technical details (they are similar to the proof in integration theory that the integral of step functions does not depend on the representation as a step function).

It is clear that ω_X is an algebra homomorphism and it remains to show that ω_X is isometric on $\mathrm{lin}\,\{\chi_O \mid O \subset \Omega_X \text{ clopen}\}$ (this will complete the proof since the unique continuous extension of an isometric algebra homomorphism is also an isometric algebra homomorphism; note that the range of the extension is $C(X)$ since $\mathbb{P}_L$ is in this range and $\mathbb{P}_L$ generates $C(X)$).

If f is any function in $\mathrm{lin}\,\{\chi_O \mid O \subset \Omega_X \text{ clopen}\}$, we choose disjoint nonvoid clopen sets $O_1,\dots,O_n$ and scalars $a_1,\dots,a_n$ such that $f = \sum_i a_i\, \chi_{O_i}$ (the case $f = 0$ is trivial so that we may assume that $f \neq 0$). With $E_i := \omega_X(O_i)$ $(i=1,\dots,n)$ the $E_1,\dots,E_n$ are disjoint (i.e. $E_iE_j = 0$ if $i \neq j$) and non-zero.

For $x \in X$ we have

$$\begin{aligned} \|\sum_i a_iE_ix\| &= \sum_i \|a_iE_ix\| \\ &\le (\max_i |a_i|)(\sum_i \|E_ix\|) \\ &\le (\max_i |a_i|)\, \|x\| \end{aligned}$$

so that $\|\sum_i a_iE_i\| \le \max_i |a_i|$

(we used the fact that, for $y_i \in \text{range } E_i$, $\|\sum_i y_i\| = \|E_1 \sum_i y_i\| + \|(Id-E_1) \sum_i y_i\| = \|y_1\| + \|y_2+\dots+y_n\| = \dots = \|y_1\| + \dots + \|y_n\|$ and that $\|x\| = \|E_1x\| + \|(Id-E_1)x\| = \|E_1x\| + \|E_2(Id-E_1)x\| +$

$\| (Id-E_2)(Id-E_1)x \| = \| E_1x \| + \|E_2x\| + \|x-E_1x-E_2x\| = \ldots =$

$\| E_1x \| + \ldots + \|E_nx\| + \| x-E_1x-\ldots-E_nx \| \geq \|E_1x\| + \ldots + \| E_nx \|)$.

Conversely, choose $x_{i_o} \in$ range E_{i_o}, $\|x_{i_o}\| = 1$ for every $i_o \in \{1,\ldots,n\}$.

Then $\|\sum_i a_iE_ix_{i_o}\| = |a_{i_o}| \, \|E_{i_o}x_{i_o}\| = |a_{i_o}|$ so that $\|\sum_i a_iE_i\| \geq \max_i |a_i|$.

This proves that
$$\|\omega_X(f)\| = \|\sum_i a_iE_i\| = \max_i |a_i| = \|f\| \quad .$$
□

<u>1.17 Corollary</u>: The idempotents in $C(X)$ are just the operators in $\mathbb{P}_L$

<u>Proof</u>: The idempotents in $C(\Omega_X)$ are just the characteristic functions of clopen subsets of Ω_X .
□

For <u>further properties of $C(X)$</u> we refer the reader to <u>chapter 3 and chapter 6</u>.

The similar construction for $\mathbb{P}_M$ instead of $\mathbb{P}_L$ will be of less importance for our considerations.

By theorem 0.5 there is a totally disconnected compact Hausdorff space Ω_X^∞ such that the elements of $\mathbb{P}_M$ correspond to the clopen subsets of Ω_X^∞

(for example, if $X = c$, then $\Omega_X^\infty \cong \alpha\mathbb{N}$).

As in proposition 1.16 there is an isometric algebra isomorphism from $C(\Omega_X^\infty)$ onto $C_\infty(X)$, the Banach algebra generated by $\mathbb{P}_M$, and the idempotents in $C_\infty(X)$ are just the operators in $\mathbb{P}_M$ (the proofs of these facts are similar to the corresponding proofs for $C(X)$).

<u>F. Summands in subspaces and quotients</u>

The following proposition is included to complete our more or less systematic study of the general properties of L-summands, L-projections, M-summands, and M-projections:-

1.18 Proposition: Let X be a Banach space and J an L-summand (M-summand) of X.

(i) For $E \in \mathbb{P}_L(X)$ ($E \in \mathbb{P}_M(X)$) we have $E(J) \subset J$. The mapping $E_J : J \to J$, $x \mapsto Ex$, is an element of $\mathbb{P}_L(J)$ (of $\mathbb{P}_M(J)$), and $E \mapsto E_J$ maps $\mathbb{P}_L(X)$ onto $\mathbb{P}_L(J)$ ($\mathbb{P}_M(X)$ onto $\mathbb{P}_M(J)$)

(ii) The L-summands (M-summands) in J are just the L-summands (M-summands) of X contained in J

(iii) By (i), every $E \in \mathbb{P}_L(X)$ ($E \in \mathbb{P}_M(X)$) induces a map $\tilde{E}: X/J \to X/J$. $E \mapsto \tilde{E}$ maps $\mathbb{P}_L(X)$ onto $\mathbb{P}_L(X/J)$ ($\mathbb{P}_M(X)$ onto $\mathbb{P}_M(X/J)$)

(iv) The L-summands (M-summands) in X/J are just the canonical images of the L-summands (M-summands) of X

Proof: We indicate the proof for the case of L-projections. The proofs for M-projections are similar.

(i) Let E_o be the L-projection whose range is J. $EE_o = E_oE$ implies that $E(J) \subset J$.

It is clear that E_J is a projection which satisfies the norm condition for L-projections so that $E_J \in \mathbb{P}_L(J)$. Conversely, for $\hat{E} \in \mathbb{P}_L(J)$, define $E: X \to X$ by $x \mapsto \hat{E}E_ox$. It is easy to see that E is an L-projection with $E_J = \hat{E}$.

(ii) follows at once from (i)

(iii) and (iv) can be reduced to (i) and (ii) since X/J is isometrically isomorphic to $J^{\perp}$. □

2. M-Ideals

We recall that we are looking for a theory which, for given Banach spaces X, would enable us to determine to what extent X behaves like a CK-space. Up to now, our only candidates for the "M-structure" of X are the M-summands of X. However, CK-spaces (cf. example 2 on p. 10) need not have any nontrivial M-summands. This fact necessitates a less restrictive definition than that of M-summands.

Before treating this problem in the present chapter we note that the situation is completely different in the case of the dual concept of "L-structure". It can be shown that a Banach space is an abstract L-space iff it has a sufficient number of L-summands. For details we refer the reader to [29] and [16] , chapter 4.

In chapter 0, section A, we collected together some well-known facts from elementary functional analysis which make it possible to guarantee the existence of elements of X which prescribed properties provided one has sufficient information concerning the Banach space geometry of X'. Thus it is to be expected that a subspace of X will behave similarly to an M-summand if its annihilator in X' has the same geometric properties as the annihilator of an M-summand. Using proposition 1.5 we are thus led to the following weakening of the definition "M-summand": we say that a closed subspace J of X is an <u>M-ideal</u> if J^{π}, the annihilator of J in X', is an L-summand in X'.

<u>Section A</u> contains some examples and a number of consequences of the definition which are more or less corollaries to the results of chapter 1. The examples show that M-ideals are much more appropriate than M-summands to describe the "M-structure" of a space.

In view of the importance of M-ideals it is useful to prove a theorem by which M-ideals in X can be characterized without knowing the L-summands of X'. Such a characterization by intersection properties of balls will be given in section D. To prepare this we prove a characterization theorem for L-summands (section B) and investigate the relations between intersection properties of balls in X and properties of certain compact convex subsets of X' (section C).

A. The structure of the collection of M-ideals in a Banach space

2.1 Definition: Let X be a Banach space and J a closed subspace of X. J is called an M-ideal if J^{π}, the annihilator of J in X', is an L-summand of X'.

By prop. 1.5 it is clear that every M-summand is an M-ideal. The converse is false (see example 1 on p.36). The following proposition shows that this is due to the fact that the complementary L-summand of a weak*-closed L-summand must not be weak*-closed.

2.2 Proposition: Let J be an M-ideal in the Banach space X.

(i) The following are equivalent:

a) J is an M-summand

b) $(J^{\pi})^{\perp}$ is weak*-closed

c) the L-projection E onto J^{π} is weak*-continuous

d) for every $x \in X$ there is a $y \in J$ such that $p(x) = p(y)$ for every $p \in (J^{\pi})^{\perp}$

(ii) If J is reflexive, then J is an M-summand

Proof:

(i) "a ⇒ d": Let E_o be the M-projection associated with J. We have $(J^{\pi})^{\perp} = \ker (Id - E_o')$ so that, for every $x \in X$, the vector $y := E_o x$ has the claimed properties.

"d ⇒c": For $x \in X$, choose $y \in J$ as in d). Since $p - Ep \in (J^\pi)^\perp$ for every $p \in X'$ it follows that $(p - Ep)(x) = (p - Ep)(y) = p(y)$, i.e. $(Ep)(x) = p(x-y)$. This proves that E is weak*-continuous.

"c ⇒b": $(J^\pi)^\perp$ is the kernel of E

"b ⇒a": We define $J^\perp := \{x \mid x \in X,\ p(x)=0 \text{ for every } p \in (J^\pi)^\perp\}$. Since $(J^\pi)^\perp$ is weak*-closed we have $(J^\perp)^\pi = (J^\pi)^\perp$.
For $x \in J$ and $x^\perp \in J^\perp$ we may regard x and $x^\perp$ as vectors in X'' so that, by prop. 1.5, $\|x+x^\perp\| = \max\{\|x\|,\|x^\perp\|\}$ (the annihilators of J^π and $(J^\pi)^\perp$ in X'' are complementary M-summands). This norm condition implies that $J \cap J^\perp = \{0\}$ so that $J + J^\perp$ is isometrically isomorphic to the product $J \times J^\perp$, the product provided with the norm $\|(x,x^\perp)\| := \max\{\|x\|,\|x^\perp\|\}$. Since $J \times J^\perp$ is a Banach space, $J + J^\perp$ must be closed in X. $J + J^\perp$ is in fact all of X since every functional $p \in X'$ which annihilates $J + J^\perp$ lies in $J^\pi \cap (J^\perp)^\pi = J^\pi \cap (J^\pi)^\perp = \{0\}$. Hence J and $J^\perp$ are complementary M-summands.

(ii) We claim that (i)d is satisfied.
The map $\omega : (J^\pi)^\perp \to J'$, $\omega(p) := p|_J$, is an isometric isomorphism.

[It is clear that ω is one-one and onto. For $q \in J'$ choose an extension $p \in X'$ such that $\|p\| = \|q\|$ and decompose p as $p = \lambda p_1 + (1-\lambda)p_2$, where $p_1 \in J^\pi$, $p_2 \in (J^\pi)^\perp$, $\lambda \in [0,1]$, $\|p_1\| = \|p_2\| = \|p\|$. Clearly $(1-\lambda)p_2$ is an extension of q with norm less than or equal to $(1-\lambda)\|q\|$ so that, unless $q=0$, it follows that $\lambda = 0$, i.e. $p = p_2$.]

For $x \in X$ we define $e_x : J' \to \mathbb{K}$ by $e_x(q) := (\omega^{-1}(q))(x)$. We have $e_x \in J''$ so that there is a $y \in J$ such that $q(y) = (\omega^{-1}(q))(x)$ for every $q \in J'$. Thus $p(x) = e_x(p|_J) = p(y)$ for every $p \in (J^\pi)^\perp$. □

Remark: It follows from the proof of (ii) that for $q \in J'$ there is only one $p \in X'$ (namely $p = \omega^{-1}(q)$) such that $p|_J = q$ and $\|p\| = \|q\|$: M-ideals have the unique Hahn-Banach extension property.

$\{0\}$ and X are called the <u>trivial M-ideals</u> of X.

<u>Examples</u>:

1. Let K be a locally compact Hausdorff space. The M-ideals in C_oK are exactly the spaces $J_C := \{ f \mid f \in C_oK , f|_C = 0 \}$, where C is a closed subset of K.

   This example shows that generally there are a lot more M-ideals than M-summands in a Banach space. We also see that C_oK-spaces have "sufficiently many" M-ideals whereas they may not have any nontrivial M-summands at all.

   (The proof will be given on p.40).

2. More generally, if A is a B^*-algebra, then the M-ideals in A are just the closed two-sided ideals of A ([82]).

   In A_{sa}, the real Banach space of the self-adjoint elements of A, the M-ideals are precisely the self-adjoint parts of the closed two-sided ideals of A ([4], prop. 6.18).

   Note that these results contain example 5 on page 10 as a special case.

Further examples will be discussed in chapter 5 and chapter 6.

The following properties of M-ideals are easy consequences of the results of chapter 1.

<u>2.3 Proposition</u>: M-ideals are $\mathbb{R}$-determined.

More precisely: if X is a complex Banach space and J a closed subspace, then J is an M-ideal in X iff J is an M-ideal in $X_{\mathbb{R}}$

<u>Proof</u>: Let $\omega : (X')_{\mathbb{R}} \to (X_{\mathbb{R}})'$ denote the isometric isomorphism of real Banach spaces $p \mapsto \mathrm{Re}\, p$. If J is an M-ideal in X, then the annihilator of J in $(X_{\mathbb{R}})'$ is just $\omega(J^{\pi})$ (since $p|_J = 0$ iff $\mathrm{Re}\, p|_J = 0$). Thus J is an M-ideal in $X_{\mathbb{R}}$ (note that the image of an L-summand under an isometric isomorphism is also an L-summand).

Conversely, let J be an M-ideal in $X_{\mathbb{R}}$ and $J^{\pi *}$ the annihilator of J in $(X_{\mathbb{R}})'$. $\omega^{-1}(J^{\pi *})$ is an L-summand in $(X')_{\mathbb{R}}$ and therefore, by th. 1.12(i), an L-summand in X'. In particular, $\omega^{-1}(J^{\pi *})$ is a $\mathbb{C}$-linear subspace so that $\omega^{-1}(J^{\pi *}) = J^{\pi}$. This proves that J is an M-ideal in X. □

<u>2.4 Proposition</u>: Let X be a Banach space which is not isometrically isomorphic to the real Banach space l^{∞}_{2}. If X contains a nontrivial L-summand (a nontrivial M-ideal) then every M-ideal (every L-summand) of X is trivial

<u>Proof</u>: This follows at once from cor. 1.14 □

Thus, for example, spaces $L^1(S,\Sigma,\mu)$ which are at least three-dimensional have no nontrivial M-ideals. Also, we obtain a new proof of the fact that C_oK-spaces have no nontrivial L-summands if K contains more than two points.

<u>2.5 Proposition</u>: Finite intersections and arbitrary closed linear spans of M-ideals are also M-ideals

<u>Proof</u>: Let J_1, J_2 be M-ideals in X. $(J_1 \cap J_2)^{\pi}$ is the weak*-closure of $J_1^{\pi} + J_2^{\pi}$ so that, by cor. 1.8, it is sufficient to show that $J_1^{\pi} + J_2^{\pi}$ is weak*-closed.

Let S be the unit ball of X'. We claim that

$$(J_1^{\pi} + J_2^{\pi}) \cap S = \mathrm{co}[(J_1^{\pi} \cap S) \cup (J_2^{\pi} \cap S)] .$$

By E_1 and E_2 we denote the L-projections onto the L-summands J_1^{π} and J_2^{π}, respectively. Then $E := E_1 + E_2 - E_1E_2$ is the L-projection onto the L-summand $J_1^{\pi} + J_2^{\pi}$. For $x \in S$ such that $Ex = x$ we have

$$\begin{aligned} \|E_1x\| + \|E_2x - E_1E_2x\| &= \|E_1x\| + \|E_2(Id-E_1)x\| \\ &\le \|E_1x\| + \|(Id-E_1)x\| \\ &= \|x\|. \end{aligned}$$

It follows that x lies in $\mathrm{co}\,[(S \cap \mathrm{range}\, E_1) \cup (S \cap \mathrm{range}\, E_2)]$ which proves "$\subset$". The reverse inclusion is obviously valid.

Thus $(J_1^{\pi} + J_2^{\pi}) \cap S$ is weak*-closed (as the convex hull of weak*-

compact convex sets) so that, by the Krein-Šmulian theorem, $J_1^\pi + J_2^\pi$ is a weak*-closed subspace of X'. Therefore the intersection of two (and consequently of finitely many) M-ideals is an M-ideal.

Now suppose that $(J_i)_{i\in I}$ is a family of M-ideals. We have $(\overline{\text{lin}} \bigcup_{i\in I} J_i)^\pi = \bigcap_{i\in I} J_i^\pi$ so that, by th. 1.11(i), $\overline{\text{lin}} \bigcup_{i\in I} J_i$ is an M-ideal. □

<u>Notes</u>: 1. In contrast to the situation for ideals in rings arbitrary intersections of M-ideals need not be M-ideals ([4],p.138).

2. We note that this proposition guarantees the existence of a largest M-ideal in every closed subspace of X.

The preceding result implies that, for the case of two M-ideals J_1, J_2, the subspace $(J_1+J_2)^-$ is an M-ideal. We will prove at once that the sum of two M-ideals is closed so that J_1+J_2 is an M-ideal. We need the following lemma to prepare the proof of this fact.

<u>2.6 Lemma</u>: Suppose that $J_1,\dots,J_n$ are M-ideals in X, $J:=J_1\cap\dots\cap J_n$. For $x\in X$, let $[x]$ (and $[x]_i$, $i=1,\dots,n$) be the equivalence class of x in X/J (in X/J_i).
Then $\|[x]\| = \max\{\|[x]_i\| \mid i = 1,\dots,n\}$.

<u>Proof</u>: Since $J\subset J_i$ we have $\|[x]\| \geq \|[x]_i\|$ for every i.
Conversely, choose $p\in J^\pi\cap S$ (S = the unit ball of X') such that $p(x) = \|[x]\|$. Since $J^\pi\cap S = (J_1^\pi + \dots + J_n^\pi)\cap S =$ $\text{co}\,[(J_1^\pi\cap S)\cup\dots\cup(J_n^\pi\cap S)]$ (this follows by induction from $(J_1^\pi + J_2^\pi)\cap S = \text{co}\,[(J_1^\pi\cap S)\cup(J_2^\pi\cap S)]$; cf. the proof of prop. 2.5) there are $p_i\in J_i^\pi\cap S$, $\lambda_i\geq 0$ such that $\sum_{i=1}^n \lambda_i = 1$ and $p = \sum_{i=1}^n \lambda_i p_i$.
It follows that $\text{Re}\, p_{i_o}(x)\geq\|[x]\|$ for a suitable $i_o\in\{1,\dots,n\}$ so that $\|[x]_{i_o}\| = \max\{|p(x)| \mid p\in J_{i_o}^\pi\cap S\} \geq |p_{i_o}(x)| \geq \|[x]\|$.
This proves that $\|[x]\| \leq \max\{\|[x]_i\| \mid i=1,\dots,n\}$. □

<u>2.7 Proposition</u>: If $J_1,\dots,J_n$ are M-ideals in X, then $J_1+\dots+J_n$ is closed and thus an M-ideal

Proof: It suffices to consider the case of two M-ideals J_1 and J_2. The natural isomorphism from $J_1/J_1 \cap J_2$ onto $(J_1+J_2)/J_2$ is an isometry by lemma 2.6 so that $(J_1+J_2)/J_2$ is complete.
This implies that J_1+J_2 is also complete and thus closed in X (recall that Z is complete iff Y and Z/Y are complete; Z a normed linear space, Y a closed subspace of Z). □

In prop. 2.2 we investigated conditions that an M-ideal is an M-summand. Lemma 2.6 yields a further characterization.

2.8 Proposition:

(i) Let J_1 and J_2 be M-ideals in X. Then the canonical images of J_1 and J_2 in $(J_1+J_2)/(J_1 \cap J_2)$ are complementary M-summands.

(ii) If J_1 and J_2 are M-ideals in X such that $J_1 \cap J_2 = \{0\}$, then J_1 and J_2 are complementary M-summands in $J_1 + J_2$.

(iii) An M-ideal J_1 of X is an M-summand iff there is an M-ideal J_2 such that $J_1 + J_2 = X$, $J_1 \cap J_2 = \{0\}$.

Proof:

(i) Without loss of generality we may assume that $J_1 + J_2 = X$. Then $X/(J_1 \cap J_2)$ is the algebraic direct sum of the canonical images of J_1 and J_2 and it remains to show that

$$\|[x_1] + [x_2]\| = \max\{\|[x_1]\|, \|[x_2]\|\} \quad (\text{all } x_1 \in J_1,\ x_2 \in J_2).$$

Let $[x]$, $[x]_1$, $[x]_2$ denote the equivalence classes of $x \in X$ in $X/(J_1 \cap J_2)$, X/J_1, X/J_2, respectively. Lemma 2.6 gives

$$\begin{aligned}\|[x_1] + [x_2]\| &= \|[x_1+x_2]\| \\ &= \max\{\|[x_1+x_2]_1\|, \|[x_1+x_2]_2\|\} \\ &= \max\{\|[x_2]_1\|, \|[x_1]_2\|\} \\ &\le \max\{\|[x_2]\|, \|[x_1]\|\} \quad (\text{all } x_1 \in J_1,\ x_2 \in J_2).\end{aligned}$$

For arbitrary $x_1 \in J_1$ choose $p \in (J_1^\pi + J_2^\pi) \cap S$ (S = the unit ball of X') such that $\|[x_1]\| = p(x_1)$. We have $p = \lambda p_1 + (1-\lambda)p_2$ for suitable $\lambda \in [0,1]$, $p_1 \in J_1^\pi \cap S$, $p_2 \in J_2^\pi \cap S$. Since $p_1(x_1) = 0$ it follows that $(1-\lambda)p_2(x_1) = p(x_1)$ so that, for $x_2 \in J_2$,

$$\begin{aligned} \|[x_1]\| &= p(x_1) \\ &\leq p_2(x_1) \\ &= p_2(x_1 + x_2) \\ &\leq \|[x_1 + x_2]\| . \end{aligned}$$

Similarly one proves that $\|[x_2]\| \leq \|[x_1 + x_2]\|$.

(ii) and (iii) follow from (i) and the fact that M-summands are M-ideals. □

As in the case of summands the M-ideals in X determine the M-ideals in every M-ideal J of X and in the quotient X/J:-

<u>2.9 Proposition</u>: Let J be an M-ideal in X. Then

(i) The M-ideals in J are precisely the M-ideals of X which are contained in J

(ii) The M-ideals in X/J are just the canonical images of the M-ideals in X.

<u>Proof</u>:

(i) We may identify J' with the quotient X'/J^{π} so that our assertion can be derived from prop. 1.18(iv).

(ii) (X/J)' can be identified with J^{π} so that (ii) is a corollary to prop. 1.18(ii). □

We will now prove the claim made in <u>example 1</u> on p. 36 .

First, let K be a compact Hausdorff space. Every closed subset C of K induces an L-projection $E_C: \mu \rightarrow \mu|_C$ on the dual space of CK. Since the range of this mapping is just $(J_C)^{\pi}$ it follows that J_C is an M-ideal.

Conversely, let J be an M-ideal in CK. We claim that J is an ideal in the B^*-algebra CK and thus of the form J_C (C a suitable closed subset of K). For $f \in CK$ we consider the multiplication operator $M_f : CK \to CK$, $g \mapsto fg$. The transposed map $(M_f)'$ can be arbitrarily well approximated by linear combinations of L-projections E_C (since f lies in the closed linear span of $\{\chi_C \mid C \subset K,\ C \text{ closed}\}$), i.e. $(M_f)' \in C((CK)')$ (= the Cunningham algebra of $(CK)'$). In particular, $(M_f)'$ commutes with the L-projection onto J^π so that $M_f J \subset J$. Consequently, J is an ideal in CK.

The case of locally compact Hausdorff spaces K can be reduced to the preceding case. $C_o K$ is the annihilator of the closed set $\{\infty\}$ and thus an M-ideal in $C(\alpha K)$. By prop. 2.9(i) the M-ideals of $C_o K$ are those subspaces J_C for which $J_C \subset C_o K$ $(C \subset \alpha K$ closed), i.e. the subspaces $\{f \mid f \in C_o K,\ f|_C = 0\}$, where C is a closed subset of K. □

B. A characterization of L-summands

Suppose that J is a closed subspace of the Banach space X and that we have to decide whether J is an L-summand or no (in section D we will apply the results of the present section to J^π , where J is a closed subspace of X).

First we take a look at a natural candidate for $J^\perp$, the complementary L-summand. This candidate is a subset of X, and we will see that J is an L-summand iff this subset is a subspace (theorem 2.12).

If J is an L-summand with complementary L-summand $J^\perp$, then the norm condition implies that the unit ball of X is flat between $x \in J$ and $x^\perp \in J^\perp$ $(\|x\| = \|x^\perp\| = 1)$; cf. fig. 1 on p. 9 . Consequently, if $J^\perp$ is complementary to J, then $J^\perp$ has to contain all vectors $x \in X$, $\|x\| = 1$, for which $J \cap \text{face}(x) = \emptyset$ (face(x) means the smallest face of the unit ball of X which contains x).

These considerations motivate the following definition:-

2.10 Definition: Let J be a closed subspace of the Banach space X. We define $J^{(\perp)} := \{ x \mid x \in X,\ x \neq 0,\ J \cap \mathrm{face}(x/\|x\|) = \emptyset \} \cup \{0\}$

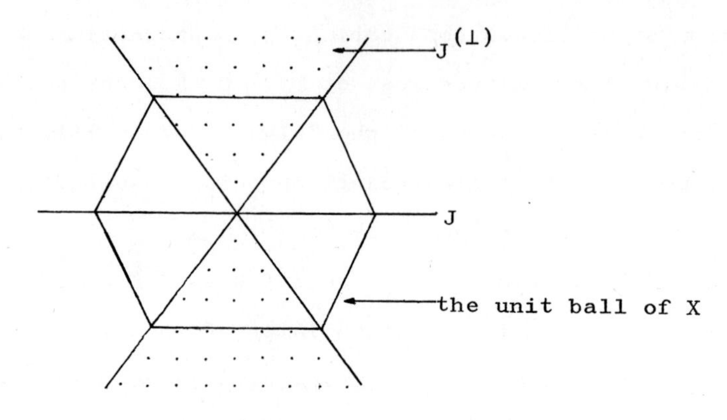

fig. 3

By the following lemma, $J^{(\perp)}$ contains sufficiently many elements to allow decompositions of arbitrary vectors such that the norm condition of def. 1.1 for L-summands is satisfied.

2.11 Lemma: For every $y_o \in X$ there are $x_o \in J$ and $x_o^{\perp} \in J^{(\perp)}$ such that

$$x_o + x_o^{\perp} = y_o \qquad \text{and} \qquad \|x_o\| + \|x_o^{\perp}\| = \|y_o\|$$

Proof: We order X by means of $\quad x \leq y \quad$ iff $\quad \|x\| + \|y-x\| = \|y\|$ (it is clear that " $\leq$ " is an order relation).

Let $A := \{x \mid x \in J,\ x \leq y_o\}$. We claim that

α) A contains a maximal element x_o

β) $x_o^{\perp} := y_o - x_o \in J^{(\perp)}$

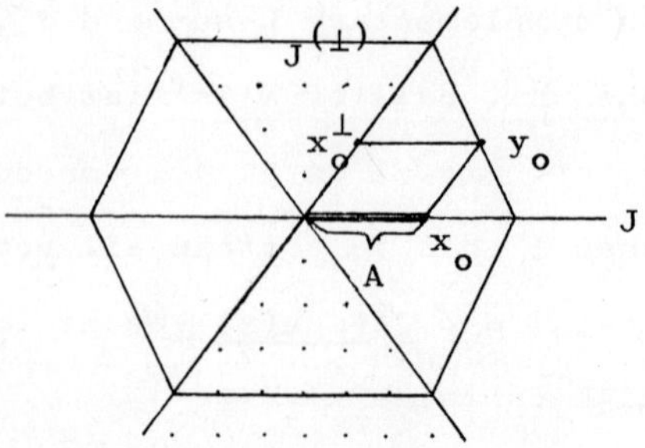

fig. 4

α) Let A' be a nonvoid totally ordered subset of A. A' is bounded in norm by $\|y_o\|$ so that $m := \sup\{\|x\| \mid x \in A'\} < \infty$.

If there is an $\tilde{x}$ in A' such that $\|\tilde{x}\| = m$, then (since A' is totally ordered) it follows that $\tilde{x}$ is the greatest element of A'.

Now suppose that $\|x\| < m$ for every $x \in A'$. We choose a sequence (x_n) in A' such that $x_1 \le x_2 \le \ldots$ and $\sup\{\|x_n\| \mid n \in \mathbb{N}\} = m$.

$\|x_{n+k} - x_n\| = \|x_{n+k}\| - \|x_n\|$ tends to zero if $k,n \to \infty$ so that $\tilde{x} := \lim x_n$ exists in J.

For $x \in A'$ we have $x \le x_n \le y_o$ if n is sufficiently large so that $x \le \tilde{x} \le y_o$. This proves that $\tilde{x}$ is an upper bound for A' in A, and Zorn's lemma provides us with a maximal element x_o of A (note that $A \neq \emptyset$ since $0 \in A$).

β) Suppose that $x_o^{\perp} := y_o - x_o \notin J^{(\perp)}$. We may assume that $\|x_o^{\perp}\| = 1$ so that $\operatorname{face}(x_o^{\perp}) \cap J \neq \emptyset$. The set

$$\left\{ x \;\middle|\; \begin{array}{l} x \in X,\ \|x\| = 1, \text{ there exist a } y \in X,\ \|y\| = 1 \text{ and a } \lambda \in \,]0,1[\, \text{ such} \\ \text{that } x_o^{\perp} = \lambda x + (1-\lambda) y \end{array} \right\}$$

is a face of the unit ball which obviously is contained in every face which contains $x_o^{\perp}$. Therefore this set is just $\operatorname{face}(x_o^{\perp})$ so that there are $x \in J$, $y \in X$, $\lambda \in]0,1[$ such that $\|x\| = \|y\| = 1$ and $x_o^{\perp} = \lambda x + (1-\lambda)y$. We have

$$\begin{aligned} 1 &= \|x_o^{\perp}\| \\ &= \|\lambda x + (1-\lambda)y\| \\ &\le \lambda\|x\| + (1-\lambda)\|y\| \\ &= 1 \end{aligned}$$

so that $\lambda x \le x_o^{\perp}$. $\lambda x \le y_o - x_o$ and $x_o \le y_o$ easily imply that $x_o + \lambda x \le y_o$ and $x_o \le x_o + \lambda x$. Since $\lambda x \neq 0$ this contradicts the maximality of x_o. □

<u>2.12 Theorem</u>: Let J be a closed subspace of X. Then the following are equivalent:

a) J is an L-summand

b) $J^{(\perp)}$ is a subspace of X

c) $J \cap \{x_1+x_2+x_3 \mid x_1,x_2,x_3 \in J^{(\perp)}\} = \{0\}$

Proof:

"a⇒b": It is easy to see that, if J is an L-summand with complementary L-summand $J^{\perp}$, we have $J^{\perp} = J^{(\perp)}$.

"b⇒c": This follows at once from $J \cap J^{(\perp)} = \{0\}$

"c⇒b": It is clear that $\mathbb{R} J^{(\perp)} \subset J^{(\perp)}$. For complex spaces X, $x \mapsto \lambda x$ is an isometric isomorphism for every λ such that $|\lambda| = 1$ which implies that face(λx) = λface(x) (for $x \in X$, $\|x\| = 1$). Consequently $\lambda J^{(\perp)} \subset J^{(\perp)}$ if $|\lambda| = 1$ so that, for arbitrary spaces, $\mathbb{K} J^{(\perp)} \subset J^{(\perp)}$.

For $x, y \in J^{(\perp)}$ we have $x + y = x_o + x_o^{\perp}$ for suitable $x_o \in J$, $x_o^{\perp} \in J^{(\perp)}$ (lemma 2.11), i.e.

$$x_o = x + y + (-x_o^{\perp}) \in J \cap \{x_1+x_2+x_3 \mid x_1, x_2, x_3 \in J^{(\perp)}\} = \{0\}.$$

Thus $x + y = x_o^{\perp} \in J^{(\perp)}$ which proves b).

"b⇒a": We have $J + J^{(\perp)} = X$ by lemma 2.11 and $J \cap J^{(\perp)} = \{0\}$ by definition. Therefore the $y \in X$ have a unique representation as $x + x^{\perp}$ ($x \in J$, $x^{\perp} \in J^{(\perp)}$), and the norm condition for L-summands is also a consequence of 2.11.

Thus, by lemma 1.2, J and $J^{(\perp)}$ are complementary L-summands. □

C. A characterization of intersection properties of balls by properties of certain compact convex subsets in the dual space

In this section we will translate intersection properties of balls into properties of sets of functionals. Our result is a generalization of the elementary fact that, for $x \in X$ and $r > 0$,

$$\|x\| < r \qquad \text{iff} \qquad \operatorname{Re} p(x) + r > 0 \text{ for every } p \in X', \|p\| \leq 1$$

2.13 Definition: Let X be a Banach space, $x \in X$ and $r \geq 0$.

$$K(x,r) := \{(p, \operatorname{Re} p(x) + r) \mid p \in X', \|p\| \leq 1\} \subset X' \times \mathbb{R}.$$

(K(x,r) is just the graph of the function

$$\operatorname{Re} p(\cdot) + r : \{p \mid p \in X', \|p\| \leq 1\} \to \mathbb{R} :$$

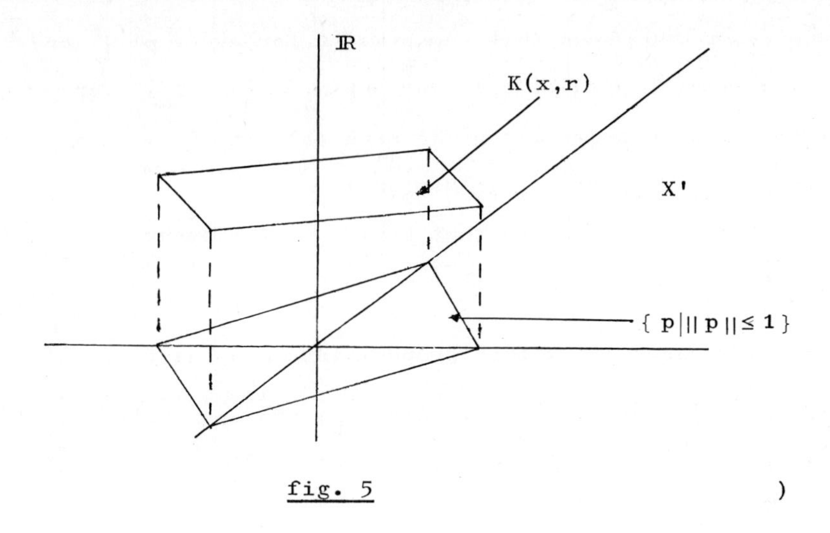

fig. 5

Since $\{ p \mid \|p\| \le 1 \}$ is weak*-compact and $p \mapsto \operatorname{Re} p(x) + r$ is weak*-continuous it is clear that $K(x,r)$ is a compact convex subset of $X' \times \mathbb{R}$ (X' provided with the weak*-topology).

<u>2.14 Proposition</u>: Let J be a closed subspace of the Banach space X, $B(x_1,r_1),\dots,B(x_n,r_n)$ a finite family of open balls ($B(x,r) := \{ y \mid y \in X, \|x - y\| < r \}$ for $x \in X$, $r > 0$).

Then $\qquad J \cap B(x_1,r_1) \cap \dots \cap B(x_n,r_n) \neq \emptyset$

$\qquad$ iff $(J^\pi \times \{0\}) \cap \operatorname{co}[K(x_1,r_1) \cup \dots \cup K(x_n,r_n)] = \emptyset$

<u>Proof</u>: Let $x \in J \cap B(x_1,r_1) \cap \dots \cap B(x_n,r_n)$ be given.

For $\sum_{i=1}^{n} \lambda_i (p_i, \operatorname{Re} p_i(x_i) + r_i) \in \operatorname{co}[K(x_1,r_1) \cup \dots \cup K(x_n,r_n)]$ it follows that $\sum_i \lambda_i (\operatorname{Re} p_i(x_i) + r_i) > \sum_i \lambda_i \operatorname{Re} p_i(x)$

$$= \operatorname{Re} \left(\sum_i \lambda_i p_i\right)(x)$$

so that, if $\sum_i \lambda_i p_i \in J^\pi$, we have $\sum_i \lambda_i (\operatorname{Re} p(x_i) + r_i) > 0$ which implies that $(J^\pi \times \{0\}) \cap \operatorname{co}[K(x_1,r_1) \cup \dots \cup K(x_n,r_n)] = \emptyset$.

Conversely, suppose that this intersection is void. Theorem 0.1 (we apply this theorem with $K_1 := J^\pi \times \{0\}$, $K_2 := \operatorname{co}[K(x_1,r_1) \cup \dots \cup K(x_n,r_n)]$; note that the convex hull of finite unions of convex compact sets is compact) provides us with an $x \in X$

and numbers $a, r > 0$ such that $\operatorname{Re} p(x) < r$ for every $p \in J^{\pi}$ and $0 < r < \operatorname{Re} p(x) + a(\operatorname{Re} p(x_i) + r_i) = a[\operatorname{Re} p(x_i + \frac{x}{a}) + r_i]$ for every $i \in \{1,\ldots,n\}$ and every $p \in X'$ such that $\|p\| \leq 1$. Thus

$$-x/a \in J \cap B(x_1,r_1) \cap \ldots \cap B(x_n,r_n)$$

(note that J^{π} is a subspace so that $p(x) = 0$ for every $p \in J^{\pi}$ and consequently $x \in J$). □

For the special case $J = X$ this proposition reads as follows:

<u>2.15 Corollary</u>: If $B(x_1,r_1), \ldots, B(x_n,r_n)$ are open balls in X, then $B(x_1,r_1) \cap \ldots \cap B(x_n,r_n) \neq \emptyset$ iff $(0,0) \notin \operatorname{co}[K(x_1,r_1) \cup \ldots \cup K(x_n,r_n)]$ □

D. A characterization of M-ideals by intersection properties

<u>2.16 Definition</u>: Let J be a subspace of the Banach space X and $n \in \mathbb{N}$.

(i) We say that J has the <u>n-ball property for open balls</u> if for every family $B(x_1,r_1),\ldots,B(x_n,r_n)$ of open balls such that $\bigcap_{i=1}^{n} B(x_i,r_i) \neq \emptyset$ and $J \cap B(x_i,r_i) \neq \emptyset$ (all $i \in \{1,\ldots,n\}$) the intersection $J \cap \bigcap_{i=1}^{n} B(x_i,r_i)$ is nonvoid

(ii) J is said to have the <u>n-ball property for closed balls</u> if $J \cap \bigcap_{i=1}^{n} D(x_i,r_i) \neq \emptyset$ for every family $D(x_1,r_1),\ldots,D(x_n,r_n)$ of closed balls ($D(x,r) := \{ y \mid y \in X, \|x-y\| \leq r\}$ for $x \in X$ and $r \geq 0$) such that $(\bigcap_{i=1}^{n} D(x_i,r_i))^{o} \neq \emptyset$ and $J \cap D(x_i,r_i) \neq \emptyset$ (all $i \in \{1,\ldots,n\}$)

It is clear that the question whether or not a subspace has such an intersection property depends on the geometry of the intersections of the unit ball of X with the translates of J. Roughly speaking, J can have an intersection property only if these intersections are "large" (if they are nonvoid).

As examples, consider $J := \{(x,0) \mid x \in \mathbb{R}\}$ in the space $\mathbb{R}^2$, where $\mathbb{R}^2$ is provided with the norm $\|(x,y)\|_1 := |x| + |y|$ (fig. 6) and the supremum norm (fig. 7):

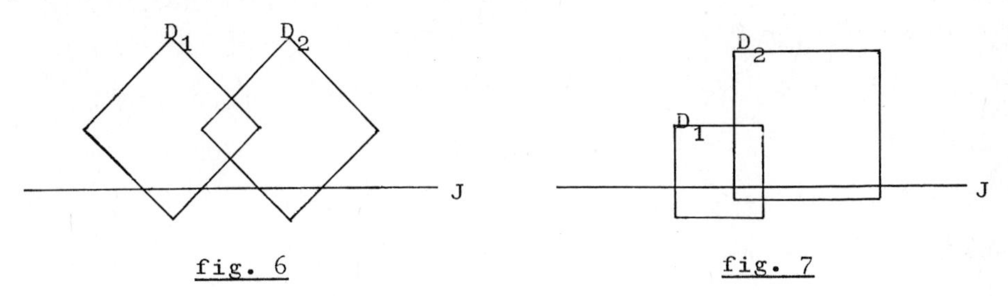

fig. 6 fig. 7

In the first case, J does not even have the two-ball property for open or closed balls whereas in the second case J has the n-ball property for open balls and for closed balls for every $n \in \mathbb{N}$ (this is an easy exercise).

We are now going to characterize M-ideals by these intersection properties. It turns out that it is sufficient to consider the case of open balls (theorem 2.17) since, as we will see in prop.2.19, J satisfies 2.16(i) for every n iff it satisfies 2.16(ii) for every n. The proof of prop. 2.19 is prepared in prop. 2.18 which states that the intersection of a finite number of balls does not grow too rapidly if these balls are "blown up". Theorem 2.20, the summary of our characterization results, contains the statements of th. 2.17 together with some equivalent statements which can easily be obtained.

2.17 Theorem: Let J be a closed subspace of the Banach space X. Then the following are equivalent:

a) J is an M-ideal

b) J has the 3-ball property for open balls

c) J has the n-ball property for open balls for every $n \in \mathbb{N}$

Proof:

"a ⇒ c": Let $B_i := B(x_i, r_i)$ be open balls such that there are $\tilde{x}_i \in J \cap B_i$ $(i=1,\dots,n)$ and $x \in \bigcap_{i=1}^{n} B_i$. We claim that

$$(J^{\pi}\times\{0\})\cap co[K(x_1,r_1)\cup\ldots\cup K(x_n,r_n)] = \emptyset$$

so that $J\cap\bigcap_{i=1}^{n} B_i \neq \emptyset$ by prop. 2.14. Suppose that there is a

$$(p,0)\in co[K(x_1,r_1)\cup\ldots\cup K(x_n,r_n)] \qquad (\text{where } p\in J^{\pi}).$$

We write $(p,0) = \sum_{i=1}^{n} \lambda_i(p_i, \operatorname{Re} p_i(x_i) + r_i)$ with $\lambda_i \geq 0$, $\| p_i \| \leq 1$, $\Sigma\ \lambda_i = 1$.

Since J^{π} is an L-summand, there are $p_i^1\in J^{\pi}$, $p_i^2\in (J^{\pi})^{\perp}$, $\mu_i\in[0,1]$ such that $p_i = \mu_i p_i^1 + (1-\mu_i)p_i^2$ and $\|p_i^1\|, \|p_i^2\|\leq\|p_i\| \leq 1$. X' is the direct sum of J^{π} and $(J^{\pi})^{\perp}$ so that (since $p = \sum_i \lambda_i\mu_i p_i^1 + \sum_i \lambda_i(1-\mu_i)p_i^2$) $p = \sum_i \lambda_i\mu_i p_i^1$ and $0 = \sum_i \lambda_i(1-\mu_i)p_i^2$. It follows that

$$\begin{aligned} 0 &= \sum_i \lambda_i(\operatorname{Re} p_i(x_i) + r_i)\\ &= \sum_i \lambda_i\mu_i(\operatorname{Re} p_i^1(x_i)+r_i) + \sum_i \lambda_i(1-\mu_i)(\operatorname{Re} p_i^2(x_i)+r_i)\\ &> \sum_i \lambda_i\mu_i \operatorname{Re} p_i^1(\tilde{x}_i) + \sum_i \lambda_i(1-\mu_i) \operatorname{Re} p_i^2(x) \end{aligned}$$

(note that $\operatorname{Re} p_i^1(x_i)+r_i > \operatorname{Re} p_i^1(\tilde{x}_i)$ since $\|x_i - \tilde{x}_i\| < r_i$, that $\operatorname{Re} p_i^2(x_i)+r_i > \operatorname{Re} p_i^2(x)$ since $\|x_i - x\|<r_i$, and that $\Sigma\lambda_i\mu_i + \Sigma\ \lambda_i(1-\mu_i) = 1$ which guarantees that there is an index i for which $\lambda_i\mu_i > 0$ or $\lambda_i(1-\mu_i) > 0$)

$$\begin{aligned} &= \sum_i \lambda_i(1-\mu_i) \operatorname{Re} p_i^2(x) \qquad (\tilde{x}_i\in J \text{ so that } p_i^1(\tilde{x}_i) = 0)\\ &= \operatorname{Re}\Big(\sum_i \lambda_i(1-\mu_i)p_i^2\Big)(x)\\ &= 0. \end{aligned}$$

This contradiction proves that

$$(J^{\pi}\times \{0\})\cap co[K(x_1,r_1)\cup\ldots\cup K(x_n,r_n)] = \emptyset .$$

"c ⇒ b": This is trivial

"b ⇒ a": We claim that condition c) of th. 2.12 is satisfied for J^{π}. Let $p_1,p_2,p_3\in(J^{\pi})^{(\perp)}$ be given and suppose that we have shown that

(*) for $x\in X$ and $\varepsilon > 0$ there is a $y\in J$ such that $\operatorname{Re} p_i(x + y) \leq \varepsilon \|p_i\|$ $(i = 1,2,3)$.

Then, if $p_1+p_2+p_3\in J^{\pi}$, it follows that (for every x and every ε and with y as in (*))

$$\operatorname{Re}(p_1+p_2+p_3)(x) = \operatorname{Re}(p_1+p_2+p_3)(x+y)$$
$$\leq \varepsilon(\|p_1\| + \|p_2\| + \|p_3\|)$$

so that $p_1+p_2+p_3$ is the zero functional.

It remains to show that (*) is valid (without loss of generality we may assume that $\|p_i\| = 1$ for $i=1,2,3$).

For $x \in X$, $\varepsilon > 0$ and $i \in \{1,2,3\}$ we consider the compact convex sets (in $X' \times \mathbb{R}$, X' provided with the weak*-topology)

$$K_1 := \operatorname{co}[\{(p, \operatorname{Re} p(x)) \mid p \in J^\pi, \|p\| \leq 1\} \cup K(0,0)]$$

and $K_2 := \{(p_i, \varepsilon)\}$:

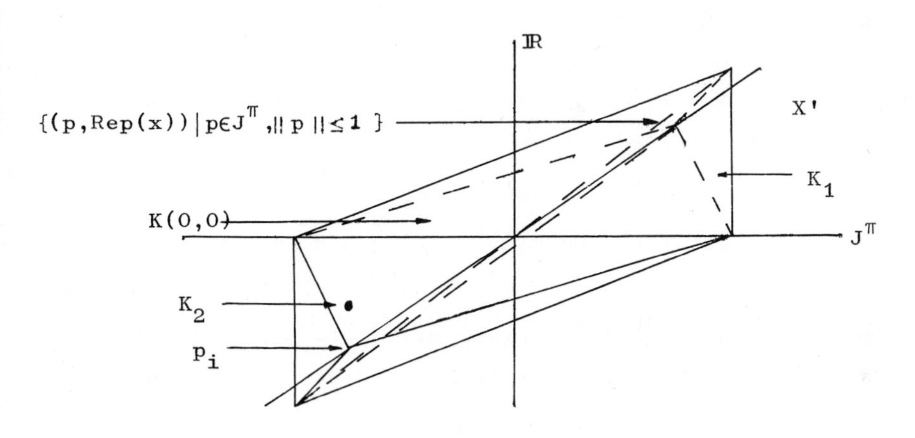

fig. 8

We claim that $K_1 \cap K_2 = \emptyset$. Otherwise, $(p_i, \varepsilon) = \lambda(q_1, \operatorname{Re} q_1(x)) + (1-\lambda)(q_2, 0)$ for some $\lambda \in [0,1]$, $\|q_1\|, \|q_2\| \leq 1$, $q_1 \in J^\pi$.

$\varepsilon = \lambda \operatorname{Re} q_1(x)$ implies that $\lambda > 0$, and $p_i = \lambda q_1 + (1-\lambda)q_2$ yields $q_1 \in \operatorname{face}(p_i)$ in contradiction to $p_i \in (J^\pi)^{(\perp)}$.

Theorem 0.1 provides us with $x_i \in X$, $r_i, a_i > 0$ such that

$\operatorname{Re} p_i(x_i) + a_i \varepsilon > r_i > \operatorname{Re} p(x_i)$, $\operatorname{Re} q(x_i) + a_i \operatorname{Re} q(x)$ (all $p \in X'$ such that $\|p\| \leq 1$, all $q \in J^\pi$ with $\|q\| \leq 1$).

With $\tilde{x}_i := -x_i/a_i$, $\tilde{r}_i := r_i/a_i$ we have $\operatorname{Re} p_i(\tilde{x}_i) + \tilde{r}_i < \varepsilon$, $\tilde{r}_i > \operatorname{Re} p(-\tilde{x}_i)$, $\operatorname{Re} q(x-\tilde{x}_i)$ (for all $\|p\|, \|q\| \leq 1$, $q \in J^\pi$). It follows that $\|\tilde{x}_i\| < \tilde{r}_i$ (which implies that $-x \in B(\tilde{x}_i - x, \tilde{r}_i)$) and

$(J^{\pi} \times \{0\}) \cap K(\tilde{x}_i - x, \tilde{r}_i) = \emptyset$ so that, by prop. 2.14, $J \cap B(\tilde{x}_i - x, \tilde{r}_i) \neq \emptyset$.

Suppose that the $B(\tilde{x}_i - x, \tilde{r}_i)$ have been constructed for $i=1,2,3$.

J has the three-ball property for open balls so that, since $-x \in \bigcap_{i=1}^{3} B(x_i - \tilde{x}, \tilde{r}_i)$ and $J \cap B(\tilde{x}_i - x, \tilde{r}_i) \neq \emptyset$ $(i=1,2,3)$, we may choose a y in $J \cap \bigcap_{i=1}^{3} B(\tilde{x}_i - x, \tilde{r}_i)$.

It follows that, for $i=1,2,3$,

$$\begin{aligned} \tilde{r}_i &> \|(\tilde{x}_i - x) - y\| \\ &\geq \operatorname{Re}(-p_i((\tilde{x}_i - x) - y)) \\ &= \operatorname{Re} p_i(x+y) - \operatorname{Re} p_i(\tilde{x}_i) \end{aligned}$$

and consequently $\operatorname{Re} p_i(x+y) \leq \operatorname{Re} p_i(\tilde{x}_i) + \tilde{r}_i \leq \varepsilon$. □

<u>2.18 Proposition</u>: Let $B(x_i, r_i)$, $i=1,\ldots,n$, be a family of n open balls in the Banach space X such that there is an x_o in $\bigcap_{i=1}^{n} B(x_i, r_i)$. Then there exists a δ in $]0,1[$ such that, for ε in $]0,1]$ and x in $\bigcap_{i=1}^{n} B(x_i, r_i + \varepsilon)$, the intersection $B(x, \delta\varepsilon) \cap \bigcap_{i=1}^{n} B(x_i, r_i + \delta\varepsilon)$ is non-empty

(we may take $\delta := (1 + \frac{m}{2M+1})^{-1}$, where $M := \max\{r_i \mid i=1,\ldots,n\}$, $m := \min\{r_i - \|x_i - x_o\| \mid i=1,\ldots,n\}$)

<u>Proof</u>: Let $\varepsilon \in \,]0,1]$ and $x \in \bigcap_{i=1}^{n} B(x_i, r_i + \varepsilon)$ be arbitrary and δ as in the proposition. We have

$$\frac{\delta}{2M+1} = \frac{1-\delta}{m} > \frac{1-\delta}{m+\varepsilon}$$

so that there is a $\lambda \in \,]0,1[$ such that $\varepsilon \frac{1-\delta}{m+\varepsilon} < \lambda < \varepsilon \frac{\delta}{2M+1}$.

This inequality implies that

$$\lambda(r_i - m) + (1-\lambda)(r_i + \varepsilon) \leq r_i + \delta\varepsilon \quad \text{and that} \quad \lambda(2M+1) < \delta\varepsilon \quad .$$

With $y := \lambda x_o + (1-\lambda)x$ it follows that

$$\begin{aligned} \|y - x\| &= \lambda \|x_o - x\| \\ &\leq \lambda(\|x_o - x_i\| + \|x_i - x\|) \\ &\leq \lambda(2r_i + \varepsilon) \\ &\leq \lambda(2M + 1) \\ &< \delta\varepsilon \end{aligned}$$

and that $\|y - x_i\| = \|\lambda x_o + (1-\lambda)x - x_i\|$

$$= \|\lambda(x_o - x_i) + (1-\lambda)(x - x_i)\|$$

$$\leq \lambda\|x_o - x_i\| + (1-\lambda)\|x - x_i\|$$

$$< \lambda(r_i - m) + (1-\lambda)(r_i + \varepsilon)$$

$$\leq r_i + \delta\varepsilon$$

so that $y \in B(x, \delta\varepsilon) \cap \bigcap_{i=1}^{n} B(x_i, r_i + \delta\varepsilon)$. □

2.19 Proposition: Let J be a closed subspace of the Banach space X and $n \in \mathbb{N}$.

(i) If J satisfies the (n+1)-ball property for open balls, then J satisfies the n-ball property for closed balls

(ii) The n-ball property for closed balls implies the n-ball property for open balls

Proof:

(i) Let $D_i := D(x_i, r_i)$ be n closed balls such that $(\bigcap_{i=1}^{n} D_i)^o \neq \emptyset$, $J \cap D_i \neq \emptyset$ for every $i \in \{1,\ldots,n\}$. We have to show that $J \cap (\bigcap_{i=1}^{n} D_i) \neq \emptyset$.

The balls $B(x_i, r_i)$ satisfy $\bigcap_{i=1}^{n} B(x_i, r_i) \neq \emptyset$ so that we may choose δ as in prop. 2.18.

$\bigcap_i B(x_i, r_i+1) \neq \emptyset$ and $J \cap B(x_i, r_i+1) \neq \emptyset$ so that there is a y_1 in $J \cap \bigcap_i B(x_i, r_i+1)$. By prop. 2.18 we have $B(y_1, \delta) \cap \bigcap_i B(x_i, r_i+\delta) \neq \emptyset$, and the (n+1)-ball property for open balls guarantees the existence of a y_2 in $J \cap B(y_1, \delta) \cap \bigcap_i B(x_i, r_i+\delta)$.

By induction we obtain a sequence $y_1, y_2, \ldots$ in J for which $y_k \in B(x_i, r_i + \delta^{k-1})$ $(i=1,\ldots,n)$, $\|y_{k+1} - y_k\| \leq \delta^k$ for every k. Thus $(y_k)_{k \in \mathbb{N}}$ is a Cauchy sequence in J, and it is clear that $\lim y_k \in J \cap \bigcap_{i=1}^{n} D_i$.

(ii) Let $B(x_i, r_i)$ be n open balls such that $\bigcap_{i=1}^{n} B(x_i, r_i) \neq \emptyset$, $J \cap B(x_i, r_i) \neq \emptyset$ for $i=1,\ldots,n$. For suitable $0 < r_i' < r_i$ we have $(\bigcap_{i=1}^{n} D(x_i, r_i'))^o \neq \emptyset$, $J \cap D(x_i, r_i') \neq \emptyset$ so that $J \cap \bigcap_{i=1}^{n} B(x_i, r_i) \supset J \cap \bigcap_{i=1}^{n} D(x_i, r_i') \neq \emptyset$. □

<u>2.20 Theorem</u>: Let J be a closed subspace of the Banach space X and $\nu: X \to X/J$ the canonical mapping onto the quotient.

The following are equivalent:

a) J is an M-ideal

b) J satisfies the 3-ball property for open balls

c) J satisfies the n-ball property for open balls (all $n \in \mathbb{N}$)

d) J satisfies the 3-ball property for closed balls

e) J satisfies the n-ball property for closed balls (all $n \in \mathbb{N}$)

f) if $B_1,\ldots,B_n$ are open balls such that $\bigcap_{i=1}^{n} B_i \neq \emptyset$, then $\nu(\bigcap_i B_i) = \bigcap_i \nu(B_i)$

g) if $D_1,\ldots,D_n$ are closed balls such that $(\bigcap_{i=1}^{n} D_i)^o \neq \emptyset$, then $\nu(\bigcap_i D_i) = \bigcap_i \nu(D_i)$

<u>Proof</u>: a) - e) are equivalent by th. 2.17 and prop. 2.19. The equivalences $c \Leftrightarrow f$ and $e \Leftrightarrow g$ are easily verified. □

<u>Note</u>: It can be shown that the two-ball property for open or closed balls does not characterize M-ideals. For a counter-example in a three-dimensional space we refer the reader to [3], p. 124.

<u>Example</u>: We already know (p. 40) that the spaces J_C are M-ideals in C_oK. As a first application of the characterization theorems we will give a simple lattice theoretical proof of this fact if K is a compact Hausdorff space and if the scalars are real.

Let C be a closed subset of K. We claim that J_C has the 3-ball property for closed balls. Let $D(f_i,r_i)$ $(f_i \in C_{\mathbb{R}}K,\ r_i > 0,\ i=1,2,3)$ be three closed balls such that there are $h \in \bigcap_{i=1}^{3} D(f_i,r_i)$ and $h_i \in J_C \cap D(f_i,r_i)$ $(i=1,2,3)$. We have

$$h_* := (f_1-\underline{r}_1) \vee (f_2-\underline{r}_2) \vee (f_3-\underline{r}_3) \leq h \leq (f_1+\underline{r}_1) \wedge (f_2+\underline{r}_2) \wedge (f_3+\underline{r}_3) =: h^* .$$

$J_C \cap D(f_i,r_i) \neq \emptyset$ implies that $h_*|_C \leq 0 \leq h^*|_C$ so that $h_o := (h^* \wedge 0) \vee h_*$ vanishes on C. It is easy to see that $h_o \in J \cap \bigcap_{i=1}^{3} D(f_i,r_i)$.

3. The centralizer

In this chapter we will investigate operators T for which the transposed operators T' behave "nicely" on the extreme functionals. Z(X), the centralizer of X, is the collection of these operators.

Centralizers of Banach spaces are fundamental to the investigations of these notes for two reasons. Firstly we will see in chapter 4 that X has a "largest" representation as a space of (vector-valued) functions, and this construction depends heavily on properties of Z(X). On the other hand the considerations of part II will show that one is often able to decide whether or not a space X has the Banach-Stone property (see def. 8.2) by looking at Z(X).

Section A is devoted to the theory of multipliers (a multiplier is an operator for which every extreme functional is an eigenvector of the transposed operator). We show that multipliers can be characterized by a certain boundedness condition. Further it is shown that structurally continuous functions on the extreme functionals define multipliers (the structure topology is constructed by means of the M-ideals of the space).

In section B we define Z(X), the centralizer of X, to be the greatest subspace of the space of multipliers which admits a natural * - operation. We consider some examples and easy consequences of the definition.

Characterization theorems for operators in the centralizer are proved in section C. We first restrict ourselves to real spaces. The theorem for the general case can be proved by reducing the problems to the underlying real spaces.

The first important applications are discussed in section D, and finally, in section E, we present a short introduction to the space of primitive M-ideals, provided with the structure topology (this is included for the sake of completeness; we prefered to investigate the structure topology on the extreme points of the dual).

A. Multipliers and M-bounded operators

It is an immediate consequence of the norm condition for L-projections that every extreme point of the unit ball is an eigenvector of such a projection (with eigenvalue 0 or 1). It follows that these extreme points are also eigenvectors for the operators in the Cunningham algebra. In particular, for operators $T:X \to X$ such that $T' \in C(X')$, every extreme functional is an eigenvector of T'. This section contains an investigation of operators with this property.

3.1 Definition: Let X be a Banach space (from now on it is tacitly assumed that $X \neq \{0\}$).
By E_X we denote the set of extreme functionals on X, i.e. the extreme points of the unit ball of X'.
An operator $T:X \to X$ is called a multiplier if every $p \in E_X$ is an eigenvector for T', i.e. if there is a function $a_T:E_X \to \mathbb{K}$ such that $p \circ T = a_T(p)p$ for every $p \in E_X$ (note that a_T is uniquely determined by T since all $p \in E_X$ are nonzero).
It is easily seen that the functions a_T are weak*-continuous and bounded (by $\|T\|$) and that $T \mapsto a_T$ is an isometric map from the set of multipliers on X (which will be denoted by Mult(X)) into the Banach algebra $m(E_X,\mathbb{K})$ of bounded $\mathbb{K}$-valued functions on E_X provided with the supremum norm.
For $S,T \in \text{Mult}(X)$, $(S\circ T - T\circ S)'$ vanishes on E_X so that $(S\circ T - T\circ S)'$ (as a weak*-continuous operator) must be the zero operator by the Krein-Milman theorem. Thus Mult(X) is a commutative operator algebra and it is obvious that it is closed in B(X). We note that $T \mapsto a_T$ is an algebra homomorphism from Mult(X) into $m(E_X,\mathbb{K})$.

[Another way of introducing Mult(X) is to consider X in the natural way as a subspace of $m(E_X, \mathbb{K})$. We say that $a \in m(E_X, \mathbb{K})$ leaves X invariant if $ax \in X$ for every $x \in X$ (more precisely: for every $x \in X$ there is a $y \in X$ such that $p(y) = a(p)p(x)$ for all $p \in E_X$). Mult(X) can be thought of as the space of these functions a .]

Examples:

1. Every operator $T: X \to X$ for which $T' \in C(X')$ is a multiplier. In particular $\mathbb{P}_M(X)$ and therefore $C_\infty(X)$ (cf. p. 31) are contained in Mult(X).
2. If K is a locally compact Hausdorff space, then $E_{C_oK} = \{\lambda \delta_k \mid \lambda \in \mathbb{K}, |\lambda| = 1, k \in K\}$; δ_k denotes the evaluation functional $f \mapsto f(k)$.
   It follows that $\mathrm{Mult}(C_oK) = \{ M_h \mid h \in C^bK\}$
   (for $T \in \mathrm{Mult}(X)$ we define $h: K \to \mathbb{K}$ by $h(k) := a_T(\delta_k)$; it is obvious that $T = M_h$ which easily implies that h is bounded and continuous; the reverse inclusion is obviously valid).
3. Let X be the complex Banach space of those continuous functions on $\{z \mid z \in \mathbb{C}, |z| \le 1\}$ which are analytic on $\{z \mid z \in \mathbb{C}, |z| < 1\}$. It can be shown that E_X is contained in the set of functionals $f \mapsto \lambda f(z)$ ($\lambda, z \in \mathbb{C}$, $|z| \le |\lambda| = 1$; see [59] , p. 145) so that M_g is a multiplier for every $g \in X$ ($M_g(f) := gf$) .

Multipliers on X can be determined without knowing E_X explicitly. By theorem 3.3 multipliers are exactly those operators which satisfy the following norm condition:-

3.2 Definition: Let T be an operator on the Banach space X. T is said to be M-bounded if there is a $\lambda > 0$ such that, for every $x \in X$,

Tx is contained in every ball which contains $\{\mu x \mid \mu \in \mathbb{K}, |\mu| \leq \lambda\}$.

Remarks:

1. We did not specify the balls in the definition to be open or closed. It is obvious that the same operators are M-bounded if we replace "balls" by "open balls" or by "closed balls".
2. Since for real space $\{\mu x \mid \mu \in \mathbb{R}, |\mu| \leq \lambda\}$ is just the convex hull of $-\lambda x, +\lambda x$ our definition agrees with the definition of M-boundedness given by Alfsen and Effros ([4], p. 150) for real spaces.
3. For $x \in X$ and $\lambda > 0$ define $R_\lambda(x) :=$

   $\bigcap\{D \mid D$ is a closed ball such that $\{\mu x \mid \mu \in \mathbb{K}, |\mu| \leq \lambda\} \subset D\}$.

   It depends on the shape of the unit ball whether $R_\lambda(x)$ is large or small. Since, for an M-bounded operator T, $Tx \in R_\lambda(x)$ for all x and a suitable λ it follows that there are nontrivial M-bounded operators only if the sets $R_\lambda(x)$ are not too small.

   For example, consider the real space $\mathbb{R}^2$, together with the norms $\|(a,b)\| := \max\{|a|,|b|\}$ (fig. 9) and $\|(a,b)\| := (a^2 + b^2)^{1/2}$ (fig. 10).

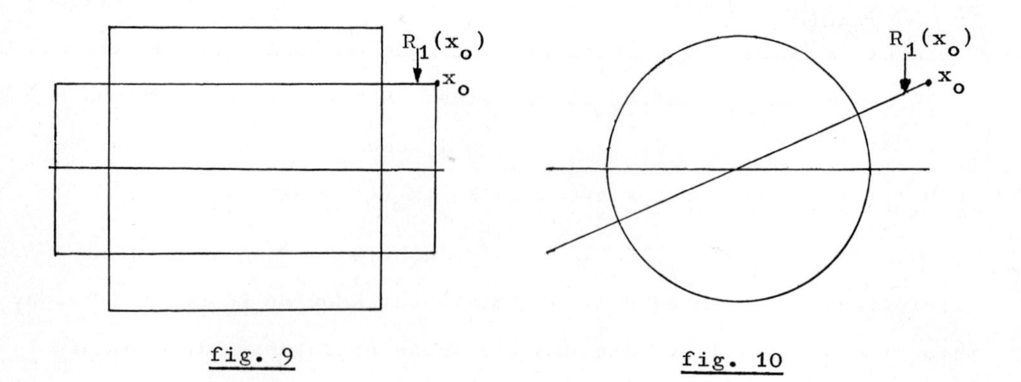

fig. 9 fig. 10

In the second case, $R_\lambda(y_o) = co\{-\lambda y_o, +\lambda y_o\}$ for every $\lambda > 0$ and every y_o so that every M-bounded operator must be a multiple of the identity operator.

It is not necessary to consider examples since, as we will see at once, an operator is a multiplier iff it is M-bounded. It is left to the reader as an exercise to show directly that the operators M_h in example 2 (p. 55) are also M-bounded operators.

3.3 Theorem: Let $T:X \to X$ be an operator. Then T is a multiplier iff T is M-bounded.

Before proving this theorem we investigate how extreme functionals behave on the sets $R_\lambda(x)$:-

3.4 Proposition: Let $x_o \in X$ and $p \in E_X$ be given such that $p(x_o) = 0$. Then $p|_{R_\lambda(x_o)} = 0$ for every $\lambda > 0$.

Proof: Since $R_\lambda(x_o) = \lambda R_1(x_o)$ it suffices to consider the case $\lambda = 1$. For $\varepsilon > 0$ choose an ε-net $\mu_1, \ldots, \mu_n$ in $\{\mu \mid \mu \in \mathbb{K}, |\mu| \leq 1\}$. Since p is an extreme point, (p, ε) is not contained in $co(\bigcup_{i=1}^{n} K(\mu_i x_o, 0))$ so that, by theorem 0.1, there is an $x \in X$ and $a, r > 0$ such that

$$\operatorname{Re} p(x) + a\varepsilon > r > \operatorname{Re}(q(x) + aq(\mu_i x_o))$$

(all $q \in X'$ such that $\|q\| \leq 1$, $i=1,\ldots,n$).

Thus, with $y := -x/a$ and $\tilde{r} := r/a$, we have $\varepsilon > \tilde{r} + \operatorname{Re} p(y)$ and $\|\mu_i x_o - y\| \leq \tilde{r}$ $(i=1,\ldots,n)$. It follows that

$$\{\mu x_o \mid |\mu| \leq 1\} \subset \{\mu_i x_o \mid i=1,\ldots,n\} + \{\mu x_o \mid |\mu| \leq \varepsilon\}$$
$$\subset D(y, \tilde{r} + \varepsilon\|x_o\|)$$

so that $R_1(x_o) \subset D(y, \tilde{r} + \varepsilon\|x_o\|)$.

For every $y_o \in R_1(x_o)$ we therefore have $\|y - y_o\| \leq \tilde{r} + \varepsilon\|x_o\|$ so that in particular $\operatorname{Re} p(y_o - y) \leq \tilde{r} + \varepsilon\|x_o\|$ and consequently

$$\operatorname{Re} p(y_o) \leq \operatorname{Re} p(y) + \tilde{r} + \varepsilon\|x_o\|$$
$$\leq \varepsilon(1 + \|x_o\|).$$

Since ε is arbitrary this implies that $\operatorname{Re} p|_{R_1(x_o)} \leq 0$, and since $\mu_o p$ is also an extreme functional for every $\mu_o \in \mathbb{K}$ with $|\mu_o| = 1$ such that $(\mu_o p)(x_o) = 0$ it follows that $\operatorname{Re} \mu_o p|_{R_1(x_o)} \leq 0$ for these μ_o. This yields $p|_{R_1(x_o)} = 0$. □

Proof of th. 3.3:

Let T be a multiplier, λ a strictly positive upper bound for $\{|a_T(p)| \mid p \in E_X\}$, and $x \in X$. Further, let $D(y,r)$ be a closed ball such that $\{\mu x \mid \mu \in \mathbb{K}, |\mu| \le \lambda\} \subset D(y,r)$.

We have to show that $\| Tx - y \| \le r$.

Let $p \in E_X$ be a functional for which $\| Tx - y\| = p(Tx - y)$. It follows that

$$\begin{aligned} \| Tx - y \| &= (p\circ T)(x) - p(y) \\ &= a_T(p)p(x) - p(y) \\ &= p(a_T(p)x - y) \\ &\le \| a_T(p)x - y\| \\ &\le r \qquad (\text{since } a_T(p)x \in \{\mu x \mid |\mu| \le \lambda\}). \end{aligned}$$

Conversely, let T be an M-bounded operator and $p \in E_X$. We have $Tx \in R_\lambda(x)$ for a suitable λ and all x so that, by prop. 3.4, $\ker p \subset \ker p\circ T$. It is a well-known elementary fact that $p\circ T$ must be a multiple of p in this case. □

Up to now, our considerations have been independent of the preceding chapters (we only noted that $C_\infty(X) \subset \mathrm{Mult}(X)$). However, we shall see at once that there is a topology on E_X, defined by means of the M-ideals of X, such that the bounded continuous functions in this topology define multipliers (th. 3.8). This topology is the structure topology on E_X. It corresponds to the Jacobson topology in ring theory (cf. section E).

3.5 Definition: Let X be a Banach space. The sets $J^\pi \cap E_X$, where J runs through all M-ideals of X, define the closed sets of a topology, the structure topology on E_X.

[note that 1. $X^\pi \cap E_X = \emptyset$ (recall that $X \ne \{0\}$)

2. $\{0\}^\pi \cap E_X = E_X$

3. $\bigcap_{i\in I} (J_i^\pi \cap E_X) = (\bigcap_{i\in I} J_i^\pi) \cap E_X = (\overline{\mathrm{lin}} \bigcup_{i\in I} J_i)^\pi \cap E_X$,

and $\overline{\mathrm{lin}} \bigcup_{i\in I} J_i$ is an M-ideal by prop. 2.5

4. If J_1, J_2 are M-ideals, then
$(J_1^\pi + J_2^\pi) \cap \{p \mid \|p\| \le 1\} =$
$[(J_1^\pi \cap \{p \mid \|p\| \le 1\}) \cup (J_2^\pi \cap \{p \mid \|p\| \le 1\})]$
(see the proof of prop. 2.5) implies that
$(J_1^\pi \cap E_X) \cup (J_2^\pi \cap E_X) = (J_1^\pi + J_2^\pi) \cap E_X$. Thus
$(J_1^\pi \cap E_X) \cup (J_2^\pi \cap E_X) = (J_1 \cap J_2)^\pi \cap E_X$, and
$J_1 \cap J_2$ is also an M-ideal by prop. 2.5]

The proof of theorem 3.8 depends on a decomposition property of M-ideals (lemma 3.6) and the fact that nontrivial M-ideals are annihilated by at least one extreme functional (lemma 3.7).

<u>3.6 Lemma</u>:

(i) If J_1 and J_2 are M-ideals, then for $x \in J_1+J_2$ and $\varepsilon > 0$ there are $x_1 \in J_1$, $x_2 \in J_2$ such that $x = x_1+x_2$ and $\|x_1\| \le (1+\varepsilon)\|x\|$

(ii) If $J_o, \ldots, J_n$ are M-ideals, then for $x \in J_o+\ldots+J_n$ and $\varepsilon > 0$ there are $x_i \in J_i$ such that $x = x_o+\ldots+x_n$ and $\|x_i\| \le (2+\varepsilon)\|x\|$ for $i=1,\ldots,n$

<u>Proof</u>:

(i) We write $x = \tilde{x}_1+\tilde{x}_2$, where $\tilde{x}_1 \in J_1$, $\tilde{x}_2 \in J_2$. The natural map from $(J_1+J_2)/J_2$ onto $J_1/J_1 \cap J_2$ is an isometrical isomorphism (see prop. 2.6) so that there is a y in $J_1 \cap J_2$ such that

$$\begin{aligned}\|\tilde{x}_1 - y\| &\le (1+\varepsilon)\|[\tilde{x}_1]_2\| \\ &= (1+\varepsilon)\|[x]_2\| \\ &\le (1+\varepsilon)\|x\|\end{aligned}$$

($[\ ]_2$ denotes the equivalence classes in $(J_1+J_2)/J_2$). Thus we may define $x_1 := \tilde{x}_1 - y \in J_1$, $x_2 := \tilde{x}_2 + y \in J_2$.

(ii) We will prove (ii) by induction on n. For n=0 there is nothing to show. Suppose that (ii) is known for a fixed number n and let $\varepsilon > 0$ and $x \in J_o+\ldots+J_{n+1}$. We choose $\tilde{\varepsilon} > 0$ such that $(2+\tilde{\varepsilon})(1+\tilde{\varepsilon}) \le 2+\varepsilon$. With $\tilde{J}_1 := J_o+\ldots+J_n$, $\tilde{J}_2 := J_{n+1}$ we may conclude from (i) that

$x = \tilde{x}_1 + \tilde{x}_2$ for suitable $\tilde{x}_1 \in \tilde{J}_1$, $\tilde{x}_2 \in \tilde{J}_2$ such that $\|\tilde{x}_1\| \leq (1+\tilde{\varepsilon})\|x\|$. By the induction assumption there are $x_o \in J_o, \dots, x_n \in J_n$ such that $\|x_i\| \leq (2+\tilde{\varepsilon})\|\tilde{x}_1\|$ $(i=0,\dots,n)$ and $\tilde{x}_1 = x_o + \dots + x_n$. With $x_{n+1} := \tilde{x}_2$ we then have $x = x_o + \dots + x_{n+1}$ and

$\|x_i\| \leq (2+\tilde{\varepsilon})\|\tilde{x}_1\| \leq (2+\tilde{\varepsilon})(1+\tilde{\varepsilon})\|x\| \leq (2+\varepsilon)\|x\|$ (for $i=0,\dots,n$),

$\|x_{n+1}\| = \|\tilde{x}_1 - x\| \leq \|\tilde{x}_1\| + \|x\| \leq (2+\tilde{\varepsilon})\|x\| \leq (2+\varepsilon)\|x\|$. □

<u>3.7 Lemma</u>:

(i) Let J be an L-summand of the Banach space X. If $J \neq \{0\}$ and if x_o is an extreme point in the unit ball of J, then x_o is also an extreme point in the unit ball of X

(ii) If J is an M-ideal of X such that $J \subsetneq X$ then there is a $p \in J^\pi \cap E_X$ (in particular we have $J = X$ iff $J^\pi \cap E_X = \emptyset$; note that $X \neq \{0\}$)

<u>Proof</u>:

(i) Let E be the L-projection onto J and suppose that $x_o = \lambda y + (1-\lambda)z$ for some $\lambda \in]0,1[$, $y,z \in X$, $\|y\|, \|z\| \leq 1$. Then $x_o = Ex_o = \lambda Ey + (1-\lambda)Ez$ so that $\|Ey\| = 1$ and therefore (since E is an L-projection) $y = Ey$. It follows that y and z are contained in the unit ball of J so that $y = z = x_o$.

(ii) J^π is isometrically isomorphic to $(X/J)'$ so that there are extreme points in the unit ball of J^π. By (i) these extreme points are contained in E_X. □

The following theorem shows that there are at least as much operators in Mult(X) as there are bounded structurally continuous functions. This result will be essential for our investigations in section C.

<u>3.8 Theorem</u>: Let $a: E_X \to \mathbb{K}$ be a bounded structurally continuous (i.e. continuous w.r.t. the structure topology) function. Then there exists a multiplier T such that $a_T = a$

<u>Proof</u>:

Suppose first that a is a real-valued structurally continuous function for which range $a \subset [0,1]$ and that $x \in X$ is an arbitrary but fixed vector.

For $n \in \mathbb{N}$ and $i \in \{0,\dots,n\}$ let $O_i :=]\frac{i-1}{n},\frac{i+1}{n}[$. Since a is continuous, there are M-ideals $J_o,\dots,J_n$ such that $J_i^{\pi} \cap E_X = \{p \mid p \in E_X,\ a(p) \notin O_i\}$. range $a \subset [0,1]$ implies $(\bigcap_{i=0}^{n} J_i^{\pi}) \cap E_X = \emptyset$ so that (by lemma 3.7) $\bigcap_{i=0}^{n} J_i^{\pi} = \{0\}$ and therefore $J_o+\dots+J_n = X$.

We choose $x_i \in J_i$ as in lemma 3.6 such that $\|x_i\| \le 3\|x\|$ $(i=0,\dots,n)$ and $x = x_o+\dots+x_n$. We define $y_n := \frac{1}{n}\sum_{i=0}^{n} i\,x_i$ and claim that

α) $|p(a(p)x - y_n)| \le \frac{6}{n}\|x\|$ for every $p \in E_X$

β) if y_n is constructed in this way for every n, then $(y_n)_{n\in\mathbb{N}}$ is a Cauchy sequence

α) Let $p \in E_X$ be arbitrary. By the construction of the O_i there is an $i_o \in \{0,\dots,n-1\}$ such that $a(p) \notin O_i$ if $i \notin \{i_o,i_o+1\}$ and $|a(p) - \frac{i}{n}| \le \frac{1}{n}$ for $i=i_o,i_o+1$. It follows that $p(x_i) = 0$ for $i \notin \{i_o,i_o+1\}$ so that

$$
\begin{aligned}
|p(a(p)x-y_n)| &= |p(\sum_{i=0}^{n}(a(p) - \tfrac{i}{n})x_i)| \\
&\le \sum_{i=0}^{n} |a(p) - \tfrac{i}{n}|\,|p(x_i)| \\
&= \sum_{i=i_o,i_o+1} |a(p) - \tfrac{i}{n}|\,|p(x_i)| \\
&\le \tfrac{1}{n}(\|x_{i_o}\| + \|x_{i_o+1}\|) \\
&\le \tfrac{6}{n}\|x\| .
\end{aligned}
$$

β) For $n,m \in \mathbb{N}$ choose $p \in E_X$ such that $\|y_n - y_m\| = |p(y_n - y_m)|$. It follows that

$$
\begin{aligned}
\|y_n-y_m\| &\le |p(a(p)x-y_n)| + |p(a(p)x-y_m)| \\
&\le (\tfrac{6}{n} + \tfrac{6}{m})\|x\| \xrightarrow[n,m\to\infty]{} 0 .
\end{aligned}
$$

We define $T:X \to X$ by $Tx := \lim y_n$. By α) we have $p(a(p)x) = p(Tx)$ for every $p \in E_X$. Thus, for $x_1,x_2 \in X$ and $\lambda \in \mathbb{K}$, $T(x_1+x_2)-Tx_1-Tx_2$ and $T(\lambda x_1) - \lambda Tx_1$ are annihilated by every $p \in E_X$ so that T is linear. T is also continuous since

$\|Tx\| = \sup\{|p(Tx)| \mid p \in E_X\} \le (\sup\{|a(p)| \mid p \in E_X\})\|x\|$ for every $x \in X$.

It is clear that T is a multiplier with $a_T = a$.

Now let a be an arbitrary bounded structurally continuous function from E_X to $\mathbb{K}$.

If $\mathbb{K} = \mathbb{R}$, choose $\mu > 0$ such that $\frac{1}{\mu}(a + \frac{\mu}{2})$ has its range in $[0,1]$. This function is also continuous so that there is a $\tilde{T} \in \mathrm{Mult}(X)$ with $a_{\tilde{T}} = \frac{1}{\mu}(a + \frac{\mu}{2})$. Obviously $T := \mu\tilde{T} - \frac{\mu}{2}\mathrm{Id}$ is also a multiplier, and $a_T = a$.

For $\mathbb{K} = \mathbb{C}$, choose $T_1, T_2 \in \mathrm{Mult}(X)$ such that $a_{T_1} = \mathrm{Re}\, a$, $a_{T_2} = \mathrm{Im}\, a$. $T := T_1 + iT_2$ has the properties claimed. □

B. The centralizer

Recall that CK-spaces are B^*-algebras so that, in order to describe the M-structure properties of arbitrary Banach spaces, we need an operation which corresponds to the map $f \mapsto \overline{f}$ for CK-spaces. The centralizer will be the set of those multipliers for which such "adjoints" can be naturally defined.

<u>3.9 Definition</u>: Let X be a Banach space

(i) For $T \in \mathrm{Mult}(X)$, we say that $S \in \mathrm{Mult}(X)$ is <u>an adjoint for T</u> if $a_S = \overline{a_T}$.
If T admits an adjoint, then this operator is clearly uniquely determined; it will be denoted by T^*

(ii) Z(X), the <u>centralizer of X</u>, is the set of those multipliers T for which an adjoint T^* exists (note that $Z(X) = \mathrm{Mult}(X)$ for real spaces)

<u>3.10 Proposition</u>:

(i) The map $T \mapsto a_T$ from Z(X) to $m(E_X, \mathbb{K})$ maps Z(X) isometrically onto a closed subalgebra of $m(E_X, \mathbb{K})$ which contains the constants and is closed under conjugation

(ii) Z(X) is a commutative B^*-algebra. Thus there is an isometric

isomorphism of B^*-algebras from $Z(X)$ onto a space CK_X, where K_X is a suitable compact Hausdorff space
(note that K_X is uniquely determined by the classical Banach-Stone theorem)

Proof:

(i) Let $\mathrm{conj}: m(E_X,\mathbb{K}) \to m(E_X,\mathbb{K})$ be the map $a \mapsto \overline{a}$ and A the closed subalgebra $\{a_T \mid T \in \mathrm{Mult}(X)\}$ of $m(E_X,\mathbb{K})$. We have $\{a_T \mid T \in Z(X)\}$ $= A \cap \mathrm{conj}^{-1}(A)$ which is closed since conj is continuous.
It is clear that $Z(X)$ is an algebra and that $\mathrm{Id} \in Z(X)$.

(ii) $T \mapsto a_T$ is an isometric isomorphism of $Z(X)$ onto a sub-B^*-algebra of $m(E_X,\mathbb{K})$. Thus $Z(X)$ is a commutative B^*-algebra with unit. □

The space K_X will be very important in what follows. For example, K_X is the largest compact Hausdorff space on which X can be represented as a space of vector-valued functions (see chapter 4).

We consider some examples:

1. We will say that X has trivial centralizer if $Z(X) = \mathbb{K}\,\mathrm{Id}$ (a number of spaces with this property will be investigated in chapter 5); K_X is the one-point space in this case
2. Let K be a locally compact Hausdorff space. Since $\mathrm{Mult}(C_oK) = \{M_h \mid h \in C^bK\}$ (cf. p. 55) and every M_h has an adjoint (namely $M_{\overline{h}}$) it follows that $Z(C_oK) = \mathrm{Mult}(C_oK) = \{M_h \mid h \in C^bK\} \cong C^bK \cong C(\beta K)$.
   This gives $K_{C_oK} = \beta K$.
3. Let A be a B^*-algebra with unit e. It can be shown ([60], p.27) that $Z(A)$ is just the space of operators $M_z: a \mapsto za$, where z runs through the elements of the centre of A.
   For A_{sa}, the real Banach space of self-adjoint elements of A, we have $Z(A_{sa}) = \{M_z \mid z$ is a self-adjoint element of the centre of A$\}$ ([4], p.167).
   Thus the centralizer can be thought of as a generalization to

arbitrary Banach spaces of the notion of centre for B^*-algebras

4. For $E \in \mathbb{P}_M(X)$ we have $E \in Z(X)$ with $E^* = E$. We will see later that every projection in $Z(X)$ is an M-projection. $\mathbb{P}_M(X) \subset Z(X)$ implies that $C_\infty(X) \subset Z(X)$.

C. Characterization theorems

In order to prepare the proof of th. 3.12 we show that operators in the Cunningham algebra satisfy a norm condition which is similar to the norm condition for L-projections.

3.11 Proposition: Let X be a real Banach space. We say that an operator T in C(X) is positive if T is contained in the closure of $\{\sum_{i=1}^{n} a_i E_i \mid a_i \geq 0,\ E_i \in \mathbb{P}_L(X),\ n \in \mathbb{N}\}$ (it is easy to see that T is positive iff the function $\omega_X^{-1}(T)$ is positive; note that Ω_X is extremally disconnected).

For $S,T \in C(X)$ we write $S \leq T$ if $T - S$ is positive. "$\leq$" is an order relation for C(X) which is compatible with the algebra structure (i.e. $S,T \geq 0$, $a \geq 0$ imply $S+T$, ST, $aT \geq 0$).

(i) If S,T are operators in C(X) such that $0 \leq S \leq T$, then

$$\|Sx\| + \|(T - S)x\| = \|Tx\| \qquad \text{for every } x \in X$$

(ii) If $T \in C(X)$ is an operator such that $0 \leq T \leq Id$, then $\lim T^n x$ exists for every $x \in X$ and $x \mapsto \lim T^n x$ is an L-projection E on X such that $0 \leq E \leq T$

Proof:

(i) For $\varepsilon > 0$ we choose L-projections $E_1,\ldots,E_n$, $\tilde{E}_1,\ldots,\tilde{E}_m$ and numbers $0 \leq a_i, b_j$ $(i=1,\ldots,n,\ j=1,\ldots,m)$ such that

$$\left\| \sum_{i=1}^{n} a_i E_i - S \right\| \leq \varepsilon, \quad \left\| \sum_{j=1}^{m} b_j \tilde{E}_j - (T - S) \right\| \leq \varepsilon.$$

We may assume that n=m, that $E_i = \tilde{E}_i$ for every i and that the E_i are disjoint (this follows from elementary properties of the Boolean algebra $\mathbb{P}_L(X)$). For $x \in X$ we have

$$\|\sum_{i=1}^{n}(a_i+b_i)E_ix\| = \sum_{i=1}^{n}(a_i+b_i)\|E_ix\| \text{ (since the } E_i \text{ are disjoint)}$$
$$= \sum_{i=1}^{n} a_i\|E_ix\| + \sum_{i=1}^{n} b_i\|E_ix\|$$
$$= \|\sum_{i=1}^{n} a_iE_ix\| + \|\sum_{i=1}^{n} b_iE_ix\|.$$

Since $\|\sum_{i=1}^{n}(a_i+b_i)E_i - T\| \le 2\varepsilon$ it follows that

$$\|Tx\| - \|Sx\| - \|(T-S)x\| = \|\sum_{i=1}^{n}(a_i+b_i)E_ix\| + \varepsilon_1 +$$
$$- \|\sum_{i=1}^{n} a_iE_ix\| + \varepsilon_2 +$$
$$- \|\sum_{i=1}^{n} b_iE_ix\| + \varepsilon_3$$
$$= \varepsilon_1 + \varepsilon_2 + \varepsilon_3 ,$$

where $|\varepsilon_1| \le 2\varepsilon\|x\|$, $|\varepsilon_2| \le \varepsilon\|x\|$, $|\varepsilon_3| \le \varepsilon\|x\|$. This proves (i).

(ii) $0 \le T \le Id$ implies that $0 \le T^n \le T^m \le Id$ for $n,m \in \mathbb{N}$, $n \ge m$. It follows that $\|T^nx\| \le \|T^mx\|$ for $n \ge m$ so that $\lim \|T^nx\|$ exists in $\mathbb{R}$ (all $x \in X$).

We have $\|T^mx-T^nx\| = \|T^mx\| - \|T^nx\|$ for $n \ge m$ so that $(T^nx)_{n\in\mathbb{N}}$ is a Cauchy sequence. We define $E:X \to X$ by $Ex := \lim T^nx$, and it remains to show that E is an L-projection.

It is clear that E is linear and that E is continuous with $\|E\| \le \|T\|$ (note that $\|T^n\| \le \|T\|$ for every n).

For $x \in X$ and $\varepsilon > 0$ we choose an $n \in \mathbb{N}$ such that $\|T^kx - Ex\| \le \varepsilon$ and $\|T^mEx - E^2x\| \le \varepsilon$ if $k,m \ge n$. Accordingly

$$\|E^2x - Ex\| \le \|E^2x-T^nEx\| + \|T^nEx-T^nT^nx\| + \|T^nT^nx-Ex\|$$
$$\le \varepsilon + \|Ex-T^nx\| + \|T^{2n}-Ex\| \text{ (since } \|T^n\|\le 1)$$
$$\le \varepsilon + \varepsilon + \varepsilon .$$

Thus E is a projection. The norm condition for L-projections follows from $\|x\| = \|T^nx\| + \|x-T^nx\|$ (since $0 \le T^n \le Id$ for every n).

Similarly $0 \le T^n \le T$ implies that $0 \le E \le T$. □

<u>3.12 Theorem</u>: Let X be a real Banach space (recall that $Z(X)$ = $Mult(X)$ in this case).

(i) If $T:X \to X$ is an operator, then the following are equivalent:

a) $T \in Z(X)$

b) T is M-bounded

c) $T' \in C(X')$

d) T is a multiplier, and a_T is structurally continuous

(ii) $\{a_T \mid T \in Z(X)\} = \left\{ a \,\middle|\, \begin{array}{l} a: E_X \to \mathbb{R} \text{ is a bounded structurally} \\ \text{continuous function} \end{array} \right\}$

<u>Proof</u>:

(ii) This follows at once from (i) and theorem 3.8

(i) a) is equivalent to b) by th. 3.3 and $d \Rightarrow a$ is trivially valid.

<u>"$c \Rightarrow d$"</u>: Suppose that $T' \in C(X')$ so that, in particular, $T \in \text{Mult}(X)$. We have to show that a_T is structurally continuous.

The following difficulty arises: We know that X' contains sufficiently many L-projections to approximate T' (so that there are "sufficiently many" L-summands in X'). However, we need M-ideals in X to guarantee the continuity of a_T, i.e. weak*-closed L-summands in X'.

At first we assume that $0 \le a_T \le 1$ and we will show that $\{p \mid p \in E_X,\ a_T(p) = 1\}$ is structurally closed.

Since $T' \in C(X')$ and $0 \le a_T$ implies that $0 \le T'$ in $C(X')$ (this is an easy consequence of the fact that the positive elements in $C(X')$ are just the squares) we get $0 \le T' \le Id$. It follows that $(T')^n$ converges pointwise to an L-projection E on X' such that $0 \le E \le T'$ (prop. 3.11(ii)).

We claim that range $E =: \tilde{J} = \{p \mid p \in X',\ p \circ T = p\}$.

"$\supset$" is trivially satisfied. Conversely, for $p \in X'$ such that $p = Ep$ the condition $0 \le E \le T' \le Id$ implies that $\|p\| \ge \|T'p\| = \|Ep\| + \|(T'-E)p\| = \|p\| + \|p \circ T - p\|$ so that $p \circ T = p$.

$\{p \mid p \in X',\ p \circ T = p\}$ is weak*-closed so that $\tilde{J}$ is a weak*-closed L-summand. Consequently $J := \{x \mid x \in X, p(x) = 0 \text{ for every } p \in \tilde{J}\}$ is an M-ideal in X such that $J^{\pi} = \tilde{J}$. Since $Ep = \lim (T')^n p = (\lim a_T(p)^n)p$ for $p \in E_X$ it is clear that $\tilde{J} \cap E_X = \{p \mid p \in E_X,\ a_T(p) = 1\}$, i.e. this set is structurally closed.

Now assume that $a_T \geq 0$ (but not necessarily ≤ 1). We choose an operator S in the closed algebra generated by T and Id such that $a_S = \inf\{\underline{1}, a_T\}$ (cf. prop. 0.2 (ii)). Clearly $S' \in C(X')$ so that (since $0 \leq a_S \leq 1$) $\{p \mid p \in E_X, a_T(p) \geq 1\} = \{p \mid p \in E_X, a_S(p) = 1\}$ is structurally closed.

It can now easily be shown (by considering $\alpha \mathrm{Id} + \beta T$ instead of T for suitable $\alpha, \beta \in \mathbb{R}$) that, for arbitrary T such that $T' \in C(X')$, the sets $\{p \mid p \in E_X, a_T(p) \leq r\}$ and $\{p \mid p \in E_X, a_T(p) \geq r\}$ are structurally closed for every $r \in \mathbb{R}$ so that a_T is structurally continuous.

<u>"a ⇒ c"</u>: We have to show that transposes of multipliers can be arbitrarily well approximated by linear combinations of L-projections so that, in particular, we have to prove that $\mathbb{P}_L(X')$ contains "sufficiently many" elements.

First, let T be a multiplier such that $0 \leq a_T \leq 1$. We claim that

(*) there exists an L-projection E in X' such that
$(Ep)(x) = \lim p(T^n x)$ for every $p \in X'$, $x \in X$.

Let $x \in X$ be a fixed vector and $p_o \in X'$ a functional such that

$$p_o \in \overline{\mathrm{co}}\{p \mid p \in E_X, p(x) \geq 0\} =: E^+(x)$$

(closure with respect to the weak*-topology). For $n \geq m$ we have $0 \leq a_{(T^n)} = (a_T)^n \leq (a_T)^m = a_{(T^m)} \leq 1$ so that

$$p_o \in \{p \mid p \in X', 0 \leq p(T^n x) \leq p(T^m x), \text{ all } n, m \in \mathbb{N}, n \geq m\}$$

(this set is weak*-closed and convex and contains all $p \in E_X$ such that $p(x) \geq 0$). It follows that $p_o(Tx) \geq p_o(T^2 x) \geq p_o(T^3 x) \geq \ldots \geq 0$ so that $\lim p_o(T^n x)$ exists.

Since the convex hull of the union of two weak*-compact convex sets is closed it follows that

$$\begin{aligned} \{p \mid p \in X', \|p\| \leq 1\} &= \overline{\mathrm{co}}\, E_X \\ &= \overline{\mathrm{co}}\, (E^+(x) \cup E^+(-x)) \\ &= \mathrm{co}\, (E^+(x) \cup -E^+(x)) \end{aligned}$$

so that X' is the linear span of $E^+(x)$. Consequently $\lim p(T^n x)$ exists for every $p \in X'$ and every $x \in X$. We define $E: X \to X$ by

$(Ep)(x) := \lim p(T^n x)$. E is well-defined since $|\lim p(T^n x)| \leq \|p\|(\sup_n \|T^n\|)\|x\| \leq \|p\|\,\|x\|$.

For $x, y \in X$ such that $\|x\|, \|y\| \leq 1$ and $n \in \mathbb{N}$ we consider $\{ p \mid p \in X', \|p\| \leq 1, p(T^n x + y - T^n y) \leq 1 \}$. This set is weak*-closed and convex and contains E_X so that it is just the unit ball of X'. Thus, for $p \in X'$ such that $\|p\| = 1$ we have $p(T^n x) + p(y) - p(T^n y) \leq 1$ and therefore $(Ep)(x) + (p - Ep)(y) \leq 1$. This yields $\|Ep\| + \|p - Ep\| \leq 1 = \|p\|$. It follows that $\|p\| = \|Ep\| + \|p - Ep\|$ for every $p \in X'$ and it remains to show that E is idempotent.

For $\varepsilon > 0$, $x \in X$, $p \in X'$ there is an $n \in \mathbb{N}$ such that $|(E^2 p)x - (Ep)(T^n x)| \leq \varepsilon$ and $|(Ep)x - p(T^k x)| \leq \varepsilon$ for $k \geq n$. Further there exists an $m \in \mathbb{N}$ such that $|(Ep)(T^n x) - p(T^m(T^n x))| \leq \varepsilon$. Thus $|(E^2 p)x - (Ep)x| \leq |(E^2 p)x - (Ep)(T^n x)| + |(Ep)(T^n x) - p(T^{m+n} x)| + |p(T^{m+n} x) - (Ep)x| \leq 3\varepsilon$. This proves that $E^2 = E$.

Now let T be an arbitrary multiplier and $\varepsilon > 0$. Suppose that we have shown that

(**) there exist multipliers $S_o, \ldots, S_k$ and real numbers $b_o, \ldots, b_k$ such that $0 \leq a_{S_i} \leq 1$ $(i = 0, \ldots, k)$ and $|a_T - \sum_{i=0}^{k} b_i (a_{S_i})^n| \leq \varepsilon$ for every $n \in \mathbb{N}$.

Then, with $\tilde{S}_n := \sum_{i=0}^{k} b_i S_i^n$, it follows from (*) that $(p(\tilde{S}_n x))_{n \in \mathbb{N}}$ converges to $(\sum_{i=0}^{k} b_i E_i)(p)(x)$ (all $p \in X'$, $x \in X$), where $\sum_{i=0}^{k} b_i E_i \in \operatorname{lin} \mathbb{P}_L(X')$.

$|a_T - \sum_{i=0}^{k} b_i (a_{S_i})^n| = |a_T - a_{\tilde{S}_n}| \leq \varepsilon$ implies that $\|T - \tilde{S}_n\| \leq \varepsilon$ so that in particular $|p(Tx) - p(\tilde{S}_n x)| \leq \varepsilon$ for $\|p\| \leq 1, \|x\| \leq 1$ (all $n \in \mathbb{N}$). Accordingly $|p(Tx) - (\sum_{i=0}^{k} b_i E_i)(p)(x)| \leq \varepsilon$ for these p,x which gives $\|T' - \sum_{i=0}^{k} b_i E_i\| \leq \varepsilon$.

For the proof of (**) we recall that closed subalgebras A of $m(E_X, \mathbb{R})$ are sublattices (cf. prop. 0.2; note that $m(E_X, \mathbb{R})$ is a CK-space). In particular, for $S_1, S_2 \in \operatorname{Mult}(X)$ there are $S^\wedge, S^\vee \in \operatorname{Mult}(X)$ such that $a_{(S^\wedge)} = \inf\{a_{S_1}, a_{S_2}\}$, $a_{(S^\vee)} = \sup\{a_{S_1}, a_{S_2}\}$.

Let $m \in \mathbb{N}$ be a number such that $\frac{1}{m} < \varepsilon$ and $0 \le a_T + m \le 2m$. For $i \in \{0,1,\dots,2m^2-1\}$ choose an operator $S_i \in \mathrm{Mult}(X)$ such that
$a_{S_i} = \inf\{\underline{1}, m\cdot\sup\{\underline{0}, a_T+m-\frac{i}{m}\}\}$.
Explicitly we have (for $p \in E_X$):

$$a_{S_i}(p) := \begin{cases} 1 & \text{if } a_T(p)+m \ge \frac{i+1}{m} \\ 0 & \text{if } a_T(p)+m \le \frac{i}{m} \\ m(a_T(p)+m-\frac{i}{m}) & \text{if } \frac{i}{m} \le (a_T(p)+m) \le \frac{i+1}{m} \end{cases} .$$

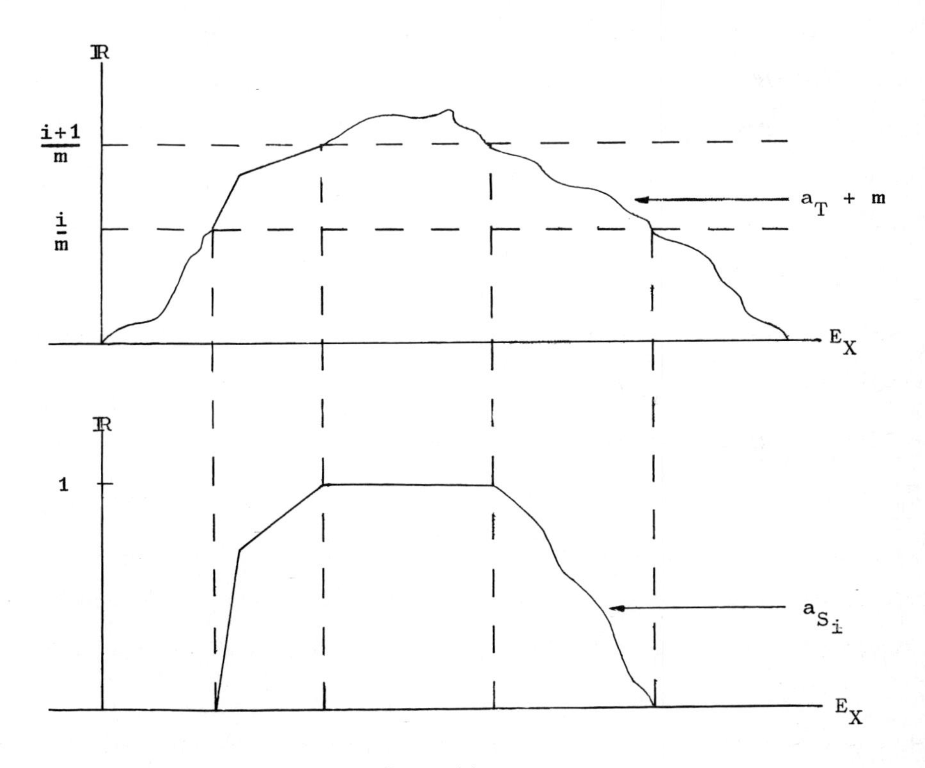

fig. 11

Using this it is easy to see that $-\varepsilon \le a_T+m-\sum_{i=0}^{2m^2-1} \frac{1}{m}(a_{S_i})^n \le \varepsilon$ for every $n \in \mathbb{N}$. With $b_0 = \dots = b_{2m^2-1} := \frac{1}{m}$, $b_{2m^2} := -m$, and $S_{2m^2} := \mathrm{Id}$ we then have $|a_T - \sum_{i=0}^{2m^2} b_i(a_{S_i})^n| \le \varepsilon$ (all $n \in \mathbb{N}$) . □

3.13 Theorem: Let X be a Banach space.

(i) If $T:X \to X$ is an operator, then the following are equivalent:

a) $T \in Z(X)$

b) T is M-bounded and T has an adjoint T^* which is also M-bounded

c) $T' \in C(X')$, and T has an adjoint T^*

d) T is a multiplier, and a_T is structurally continuous

If X is a complex Banach space, then a) - d) are also equivalent to

e) $T \in Z(X_{\mathbb{R}}) + iZ(X_{\mathbb{R}})$

(ii) $\{ a_T \mid T \in Z(X) \} = \{ a \mid a:E_X \to \mathbb{K}$ is a bounded structurally continuous function$\}$

Proof: For $\mathbb{K} = \mathbb{R}$ these are just the assertions of theorem 3.12. Therefore it suffices to consider a complex Banach space X.

(i) "a ⇔ b": This follows from th. 3.3

"d ⇒ a": If a_T is structurally continuous, then so is $\overline{a_T}$. By theorem 3.8 there exists a multiplier T^* such that $a_{T^*} = \overline{a_T}$. Thus T admits an adjoint T^*.

"a ⇒ c": We define $T_1 := \frac{T+T^*}{2}$, $T_2 := \frac{T-T^*}{2i}$. T_1 and T_2 are multipliers such that range a_{T_1} and range a_{T_2} are contained in $\mathbb{R}$ so that $T_1, T_2 \in Mult(X_{\mathbb{R}})$. By th. 3.12, a ⇔ c, $T_1', T_2' \in C((X_{\mathbb{R}})')$, and $C((X_{\mathbb{R}})') \subset C(X')$ by th. 1.12. It follows that $T' = (T_1 + iT_2)' = T_1' + iT_2' \in C(X')$ (and also that $(T^*)' \in C(X')$).

"c ⇒ e": As in the preceding part of the proof we consider $T_1 := \frac{T+T^*}{2}$, $T_2 := \frac{T-T^*}{2i}$. T_1 and T_2 are multipliers with real-valued a_{T_1}, a_{T_2}. Thus $T_1, T_2 \in Mult(X_{\mathbb{R}}) = Z(X_{\mathbb{R}})$ so that $T = T_1 + iT_2 \in Z(X_{\mathbb{R}}) + iZ(X_{\mathbb{R}})$.

"e ⇒ d": Suppose that $T = T_1 + iT_2$, where $T_1, T_2 \in Z(X_{\mathbb{R}})$. a_{T_1} and a_{T_2} are continuous on $E_{X_{\mathbb{R}}}$ by th. 3.12(ii), and the structure topologies on $E_{X_{\mathbb{R}}}$ and E_X are identical by prop. 2.3. Thus $a_T = a_{T_1} + ia_{T_2}$ is continuous as the linear combination of continuous maps.

(ii) " $\subset$ " follows from (i),a $\Rightarrow$d. Conversely, if $a:E_X \to \mathbb{C}$ is structurally continuous, choose $T, T^* \in \text{Mult}(X)$ such that $a_T = a$, $a_{T^*} = \overline{a}$ (th. 3.8). T^* is adjoint to T so that $T \in Z(X)$. □

Note: In general we have $Z(X) \subsetneqq \text{Mult}(X)$ for complex Banach spaces. For example, with X as in example 3 on p. 55, the operators M_g are multipliers for $g \in X$. It can be shown that M_g admits an adjoint iff $\overline{g}$ also belongs to X, i.e. iff g is constant.

We do not know whether (i)c) in theorem 3.13 can be replaced by

$$" \quad T' \in C(X') \quad " ,$$

i.e. whether the multipliers T for which $T' \in C(X')$ always have an adjoint.

D. Applications of the characterization theorems

By the characterization theorems of the preceding section we are able to determine $Z(X)$ without knowing X' explicitly. For example,

- if E_X and the M-ideals of X are known, one has to investigate the structurally continuous functions on E_X (or the multipliers T for which a_T is structurally continuous)
- if only E_X is known, one has to consider the multipliers which admit an adjoint
- if it is not possible to determine E_X then remember that
  for real spaces: $Z(X)$ is the space of M-bounded operators
  for complex spaces: $Z(X) = Z(X_{\mathbb{R}}) + iZ(X_{\mathbb{R}})$, and $Z(X_{\mathbb{R}})$ is the space of M-bounded operators on $X_{\mathbb{R}}$.

3.14 Proposition: Let J be an M-ideal of the Banach space X. Then J is an invariant subspace for every $T \in Z(X)$. Thus, if we regard X

as a $Z(X)$-module, every M-ideal is a sub-module

Proof: Let E be the L-projection onto J^{π}. Since $C(X')$ is commutative and $T' \in C(X')$ for $T \in Z(X)$ it follows that $T'E = ET'$ so that $T(J) \subset J$. □

3.15 Proposition: The idempotent elements of $Z(X)$ are just the M-projections of X

Proof: We already noted that $\mathbb{P}_M(X) \subset Z(X)$. Conversely, let E be a projection in $Z(X)$. Then E' is a projection in $C(X')$ and thus an L-projection by cor. 1.17. Therefore E is an M-projection (see prop. 1.5(ii)). □

3.16 Proposition:

(i) If X contains no nontrivial M-ideal, then $Z(X) = \mathbb{K}\,\mathrm{Id}$

(ii) Let X be a Banach space which is not isometrically isomorphic to the real Banach space l_2^{∞}. If X contains a nontrivial L-projection, then $Z(X)$ is trivial

Proof:

(i) The structure topology is the indiscrete topology in this case so that $Z(X) = \mathbb{K}\,\mathrm{Id}$ by th. 3.13(ii).

(ii) follows immediately from the L-M-theorem 1.13 and th. 3.13,a $\Rightarrow$ c. □

Note: The converse of (i) is not true: $Z(B(H))$ is trivial since $B(H)$ has trivial centre, but $B(H)$ contains the nontrivial M-ideal $K(H)$ (H an infinite-dimensional Hilbert space).

Finally we note that, by th. 3.13(i), we can determine $Z(X)$ if $Z(X_{\mathbb{R}})$ is known so that the centralizer is $\mathbb{R}$-determined.

E. The space of primitive M-ideals

The structure topology for rings deals with topological spaces which are collections of ideals. In this section we will show that a similar construction is possible for Banach spaces and indicate how this construction is related to the structure topology on E_X.

Let X be a Banach space. For $p \in E_X$, let J_p (the <u>primitive M-ideal</u> associated with p) be the largest M-ideal contained in ker p (cf. note 2 on p. 38). Prim(X) means the collection of all J_p for $p \in E_X$. If J is an M-ideal in X, define h(J), the <u>hull</u> of J, as the set of all $J_p \in$ Prim(X) for which $J \subset J_p$. It can be shown that the hulls h(J), J an M-ideal, form the closed sets of a topology (the <u>structure topology</u>) on Prim(X). For the proof of the fact that the union of two hulls is also a hull one has to note that $J_1 \cap J_2 \subset J_p$ implies that $J_1 \subset J_p$ or $J_2 \subset J_p$ ($p \in E_X$, J_1 and J_2 M-ideals) which is an immediate consequence of

$$(J_1 \cap J_2)^{\pi} \cap E_X = (J_1^{\pi} \cap E_X) \cup (J_2^{\pi} \cap E_X)$$

(see note 4 to def. 3.5, p. 59).

The structure topologies on Prim(X) and E_X are related as follows:

<u>3.17 Proposition</u>: Let $\rho : E_X \to$ Prim(X) be the map $p \mapsto J_p$. Then $A \subset E_X$ is closed iff there is a closed set $B \subset$ Prim(X) such that $A = \rho^{-1}(B)$ □

<u>Proof</u>: This is a restatement of the obvious fact that $p \in J^{\pi}$ iff $J \subset J_p$ (J an M-ideal, $p \in E_X$)

Because of this proposition it makes no essential difference whether one investigates the structure topology on E_X or on Prim(X). Since ρ is surjective but not injective this does not mean that Prim(X) and E_X have the same topological properties. For example, Prim(X) is always a T_o-space (for $J_{p_1} \not\subset J_{p_2}$ we have $J_{p_2} \notin h(J_{p_1}) \ni J_{p_1}$) whereas E_X never has this property.

The "translation" of theorem 3.8 into a property of Prim(X) is of some interest. Suppose that a:Prim(X) $\to \mathbb{K}$ is a bounded structurally continuous function. By prop. 3.17 $a \circ \rho$ is also continuous so that there exists a multiplier T with $a_T = a \circ \rho$. We claim that $x_p := a(J_p)x - Tx \in J_p$ for every $p \in E_X$. If not, there is an extreme point $\tilde{p}$ in the unit ball of $(J_p)^{\pi}$ ($\cong (X/J_p)'$) such that $\tilde{p}(x_p) \neq 0$. We have $\tilde{p} \in E_X$ by lemma 3.7(i), and $\tilde{p}|_{J_p} = 0$ implies that $J_p \subset J_{\tilde{p}}$. Thus $J_{\tilde{p}}$ is contained in every hull which contains J_p so that, by the continuity of a, $a(J_p) = a(J_{\tilde{p}})$ and therefore $x_p = a(J_{\tilde{p}})x - Tx$. This yields $\tilde{p}(x_p) = (a(J_{\tilde{p}}) - a_T(\tilde{p}))\tilde{p}(x) = 0$, a contradiction.

It follows that $[Tx]_p = a(J_p)[x]_p$ for every $p \in E_X$ ($[\ ]_p$ denotes the equivalence classes in X/J_p) so that we may restate theorem 3.8 by saying that structurally continuous functions on Prim(X) define operators which act as multiplication operators in the quotients X/J_p. For the special case when X = A is a B^*-algebra this is just the Dauns-Hofmann theorem ([35]).

Further properties of the structure topology on Prim(X) are discussed in [4], pp. 141-146.

4. Function modules

Let K be a nonvoid compact Hausdorff space and $(X_k)_{k\in K}$ a family of Banach spaces. The product $\prod_{k\in K} X_k$ can be thought of as a space of functions on K where the values of the functions at different points of K lie (possibly) in different spaces.

The space $\prod_{k\in K} X_k$ is in general far too large to be useful in functional analysis. We will restrict our attention to the subspace $\prod^{\infty}_{k\in K} X_k$, i.e. the space of those functions x in $\prod_{k\in K} X_k$ for which $\|x\| := \sup\{\|x(k)\| \mid k \in K\} < \infty$ (we concede that it would be more accurate to write $\|x(k)\|_k$ instead of $\|x(k)\|$; however, we will continue to use only one symbol for all norms under consideration if confusion is not likely).

Closed subspaces X of $\prod^{\infty}_{k\in K} X_k$ will be called Banach spaces of vector-valued functions on K. It is clear that certain conditions must be satisfied if we are interested in getting information concerning the Banach space X from properties of K and the X_k.

[For example, if X is any Banach space, then X can be regarded as a subspace of $m(E_X,\mathbb{K})$, i.e. as a subspace of $\prod^{\infty}_{p\in E_X} \mathbb{K}_p$ (where $\mathbb{K}_p = \mathbb{K}$ for every p). However, this does not imply that any nontrivial problems concerning X can be reduced to problems concerning $\mathbb{K}$.]

The choice of these conditions should guarantee that

(*) X inherits properties of the space CK
(in particular we are interested in M-structure properties; for example, clopen subsets of K should define M-summands of X)

(**) it is possible to "translate" properties (in particular M-structure properties) of X into properties of K and the X_k.

The considerations of this chapter will show that the following definition is appropriate:

a <u>function module</u> is a Banach space X of vector-valued functions on K such that

(i) $hx \in X$ for $x \in X$ and $h \in CK$, where $(hx)(k) := h(k)x(k)$ (i.e. X is a CK-module)

(ii) $k \mapsto \|x(k)\|$ is an upper semicontinuous function for every $x \in X$

(iii) $X_k = \{x(k) \mid x \in X\}$ for every $k \in K$

(iv) $\{k \mid k \in K, X_k \neq \{0\}\}$ is dense in K .

The most important properties of function modules are (i) and (ii). (iii) and (iv) are needed to avoid trivial situations and to assure that, in a sense, the elements of K and the X_k can be approximated by the elements of X.

It will be shown that for function modules not only (*) and (**) are valid but also

(***) every Banach space is isometrically isomorphic to a function module such that K is "maximal"; in this particular representation Z(X) has a simple form: the operators in Z(X) are just the operators $x \mapsto hx$ (where $h \in CK$) so that all properties of Z(X) (and thus, by prop. 3.15, of $\mathbb{P}_M(X)$) are consequences of the topological properties of K.

In <u>section A</u> we investigate the basic properties of function modules. In particular we show that for function modules the concepts of the theory of M-structure (M-summands, M-ideals, E_X, the centralizer) behave as indicated in (*) and (**). <u>Section B</u> is mainly devoted to the verification of (***). It turns out that the "maximal" K is just K_X (see prop. 3.10(ii)) and that the component spaces, i.e. the spaces X_k, are uniquely determined in this case.
Applications of (***) are considered in <u>section C</u>. Most of the results in this section can be proved independently of the theory of function modules, but using (***) the operators in the centralizer can be regarded as simple multiplication operators so that the

proofs are particularly simple in this setting.

A. General properties of function modules

For the sake of easy reference we repeat the definition of function modules from the preceding introduction:

4.1 Definition: A function module is a triple $(K,(X_k)_{k\in K},X)$, where K is a nonvoid compact Hausdorff space (the base space), $(X_k)_{k\in K}$ a family of Banach spaces (the component spaces), and X a closed subspace of $\prod^{\infty}_{k\in K} X_k$ such that the following conditions are satisfied:

(i) $hx\in X$ for $x\in X$ and $h\in CK$ ($(hx)(k) := h(k)x(k)$)

(ii) $k\mapsto \|x(k)\|$ is an upper semicontinuous function for every $x\in X$

(iii) $X_k = \{x(k) \mid x\in X\}$ for every $k\in K$

(iv) $\{k \mid k\in K,\ X_k \neq \{0\}\}^- = K$

Remarks: 1. Instead of "$(K,(X_k)_{k\in K},X)$ is a function module" we will often say that X is a function module in $\prod^{\infty}_{k\in K} X_k$ or (if K and the $(X_k)_{k\in K}$ are understood) that X itself is a function module.

2. The definition of function module varies slightly in the various papers where this type of space of vector-valued functions is considered. Thus little care is necessary in comparing results from different papers.

Examples:

1. Let K be a nonvoid compact Hausdorff space and X_o a Banach space ($X_o \neq \{0\}$). With $X_k := X_o$ for every k and $X :=$ $\{x \mid x:K\to X_o,\ x \text{ continuous}\}$ it is easy to see that X is a function module in $\prod^{\infty}_{k\in K} X_k$

(X is just $C(K,X_o)$).

2. More generally, let L be a nonvoid locally compact Hausdorff space and X_o a Banach space ($X_o \neq \{0\}$). Let $K := \beta L$ and $X_k := X_o$ or $X_k := \{0\}$ if $k \in L$ or $k \in K \setminus L$, respectively. Then $C_o(L,X_o)$ can be regarded in a natural way as a function module with base space K and component spaces $(X_k)_{k\in K}$.
   Similarly $C_o(L,X_o)$ can be regarded as a function module with base space αL and component spaces $X_k := X_o$ if $k \in L$ and $X_\infty := \{0\}$.
   In part II we will discuss the question whether L can be reconstructed from the Banach space geometry of $C_o(L,X_o)$, and function modules will be very useful in deciding this problem.
3. Let X be a nonzero Banach space. Then, with $K := \{1\}$ and $X_1 := X$, X is a function module in $\prod^\infty_{k\in K} X_k$.
4. If $X_1,\ldots,X_n$ are nonzero Banach spaces, then $\prod^{n\ \infty}_{i=1} X_i$ is a function module (with base space $\{1,\ldots,n\}$ and component spaces $X_1,\ldots,X_n$).
5. Let K be a nonvoid compact Hausdorff space and $(X_k)_{k\in K}$ a family of Banach spaces such that $\{k \mid k\in K,\ X_k \neq \{0\}\}$ is dense in K. Then there exists a function module in $\prod^\infty_{k\in K} X_k$, for example the space $X:=\{x \mid x \in \prod^\infty_{k\in K} X_k, \{k \mid \|x(k)\| \geq \varepsilon\}$ is finite for every $\varepsilon > o\}$.
6. Let L and K be nonvoid compact Hausdorff spaces and t a continuous map from L onto K. For $k \in K$ we define $X_k := C(t^{-1}\{k\})$, and for $f \in CL$, $\tilde{f}$ denotes the element $k \mapsto f|_{t^{-1}(\{k\})}$ of $\prod^\infty_{k\in K} X_k$.
   It is easy to see that $\{\tilde{f} \mid f \in CL\}$ is a function module in $\prod^\infty_{k\in K} X_k$ which is isometrically isomorphic to CL (a similar construction is used in [71] to prove a generalized Stone-Weierstraß theorem).

We already noted that the conditions (i) and (ii) for function modules are much more essential than the conditions (iii) and (iv).

We will see at once (prop. 4.3(iii)) that every subspace of $\prod_{k\in K}^{\infty} X_k$ for which (i) and (ii) are satisfied defines a function module if K and the X_k are suitably restricted.

4.2 Lemma: Let x be an element of $\prod_{k\in K}^{\infty} X_k$ (K and $(X_k)_{k\in K}$ as in def. 4.1) such that $k \mapsto \|x(k)\|$ is upper semicontinuous, L a nonvoid closed subset of K, and U a neighbourhood of L.

Then there exists a function $h \in CK$ such that $hx|_L = x|_L$, $h|_{K\setminus U} = 0$, and $\|hx\| = \|x|_L\| := \sup\{\|x(k)\| \mid k \in L\}$.

Proof: Since $\|x(\cdot)\|$ is upper semicontinuous and L is compact there are open neighbourhoods U_n of L contained in U such that $\sup\{\|x(k)\| \mid k \in U_n\} \leq \|x|_L\| (\sum_{i=1}^{n} 2^{-i})^{-1}$. For $n \in \mathbb{N}$, choose $h_n \in CK$ such that range $h_n \subset [0,1]$, $h_n|_L = 1$, $h_n|_{K\setminus U_n} = 0$. Then $h := \sum_{i=1}^{\infty} 2^{-i} h_i$ is a continuous function on K such that $h|_L = 1$, $h|_{K\setminus U} = 0$. It is easy to see that $\|hx\| = \|x|_L\|$. □

4.3 Proposition: (K and $(X_k)_{k\in K}$ as in def. 4.1).

Let X be a subspace of $\prod_{k\in K}^{\infty} X_k$ such that $X \neq \{0\}$ and (i) and (ii) of def. 4.1 are satisfied. Then

(i) (i) and (ii) of def. 4.1 are also valid for X^-

(ii) If L is a nonvoid closed subset of K, then $X^-|_L := \{x|_L \mid x \in X^-\}$ is closed in $\prod_{k\in L}^{\infty} X_k$

(iii) $X^-|_{\tilde{K}}$ is a function module in $\prod_{k\in\tilde{K}}^{\infty} \tilde{X}_k$, where $\tilde{X}_k := \{x(k) \mid x\in X\}^-$ for $k \in K$ and $\tilde{K} := \{k \mid k \in K, \tilde{X}_k \neq \{0\}\}^-$

(iv) If X is a function module in $\prod_{k\in K}^{\infty} X_k$ and L a nonvoid closed subset of K such that $\{k \mid k\in L, X_k \neq \{0\}\}^- = L$, then $X|_L$ is a function module in $\prod_{k\in L}^{\infty} X_k$. $X|_L$ is called the restriction of X to L

Proof:

(i) This follows immediately from $\|hx - hy\| \leq \|h\|\|x - y\|$ and

$| \|x(k)\| - \|y(k)\| | \leq \|x - y\|$ (for $x,y \in \prod^{\infty}_{k\in K} X_k$, $h \in CK$, $k \in K$).

(ii) It suffices to prove that the closure of $X|_L$ is contained in $X^-|_L$. Let $y_o \in (X|_L)^-$ be arbitrary and $(x_n|_L)_{n\in\mathbb{N}}$ a sequence in $X|_L$ such that $\sum_{n=1}^{\infty} \|x_n|_L\| < \infty$ and $y_o = \sum_{n=1}^{\infty} x_n|_L$. Choose $y_n := h_n x_n$ as in lemma 4.2 such that $y_n|_L = x_n|_L$, $\|y_n\| = \|x_n|_L\|$. Since $\sum_{n=1}^{\infty} \|y_n\| < \infty$ it follows that $y := \sum_{n=1}^{\infty} y_n$ exists in X^-, and it is clear that $y|_L = y_o$.

(iii) $\prod^{\infty}_{k\in\tilde{K}} \tilde{X}_k$ is closed in $\prod^{\infty}_{k\in\tilde{K}} X_k$ so that $(X|_{\tilde{K}})^- = X^-|_{\tilde{K}} \subset \prod^{\infty}_{k\in\tilde{K}} \tilde{X}_k$. An application of (ii) (with $L = \{k\}$) gives $\tilde{X}_k = \{x(k) | x \in X^-\}$ for every $k \in K$. The other properties of a function module follow from (i) (together with the Tietze extension theorem) or are satisfied by definition.

(iv) $X|_L$ is closed in $\prod^{\infty}_{k\in L} X_k$ by (ii). 4.1(i) (for $X|_L$) is a consequence of the Tietze extension theorem, and 4.1(ii)(iii) are trivially satisfied by $X|_L$. □

4.4 Corollary: Condition (iii) in def. 4.1 can be replaced by

(iii)' $\{x(k) | x \in X\}^- = X_k$ for every $k \in K$

Proof: By 4.3(ii), $X|_{\{k\}} = \{x(k) | x \in X\}$ is closed in X_k for every closed subspace X of $\prod^{\infty}_{k\in K} X_k$ such that 4.1(i)(ii) are valid. □

We are now going to investigate the relationship between the M-structure properties of X, the X_k, and CK ($(K,(X_k)_{k\in K},X)$ a function module). Our first theorem is a strong generalization of the well-known fact that, for compact Hausdorff spaces K,

$$E_{CK} = \{\lambda\delta_k | k \in K, \lambda \in \mathbb{K}, |\lambda| = 1\} .$$

4.5 Theorem: Let $(K,(X_k)_{k\in K},X)$ be a function module and $p \in X'$. Then $p \in E_X$ iff there is a $k \in K$ such that $X_k \neq \{0\}$ and a $p_k \in E_{X_k}$ such that $p(x) = p_k(x(k))$ for $x \in X$

Because of this fact E_X can be thought of as the disjoint union $\dot{\bigcup}\{E_{X_k} | k\in K, X_k \neq \{0\}\}$.

Proof:

1. Suppose that $p \in E_X$. We first note that

(*) for $x \in X$ and $h \in CK$ we have $hx \in R_{||h||}(x)$ (cf. remark 3 on p. 56 ; this follows at once from the definition of the norm in $\prod^{\infty}_{k\in K} X_k$) so that $p(x) = 0$ implies that $p(hx) = 0$ for these h (prop. 3.4).

We claim that α) for every $x \in X$ such that $p(x) \neq 0$ there is a $k_x \in K$ such that $p(hx) \neq 0$ for every continuous $h:K \to [0,1]$ for which $h(k_x) = 1$

β) if $x_1, x_2 \in X$ are vectors such that $p(x_1) \neq 0 \neq p(x_2)$ and $k_1 := k_{x_1}$ and $k_2 := k_{x_2}$ are determined for x_1 and x_2 as in α), then $k_1 = k_2$;
it follows that there is exactly one $k \in K$ such that $p(hx) \neq 0$ for every $x \in X$ with $p(x) \neq 0$ and every continuous $h:K \to [0,1]$ with $h(k) = 1$

γ) $x(k) = 0$ implies that $p(x) = 0$ (all $x \in X$)

δ) let the mapping $p_k : X_k \to \mathbb{K}$ be defined by $p_k(x(k)) := p(x)$; p_k is well-defined and an element of E_{X_k} .

α) Suppose that $x \in X$ and that for every $l \in K$ there is a function $h_l \in CK$ such that range $h_l \subset [0,1]$, $h_l(l) = 1$, and $p(h_l x) = 0$. We will show that $p(x) = 0$ in this case.

Since K is compact, we may choose $h_{l_1}, \ldots, h_{l_n}$ such that $h := h_{l_1} + \ldots + h_{l_n}$ is strictly positive on K, say $h \geq \varepsilon > 0$. We have $p(hx) = 0$ so that, since $\frac{\varepsilon}{h}$ is continuous, it follows from (*) that $p(\frac{\varepsilon}{h} hx) = \varepsilon\, p(x) = 0$.

β) Suppose that $k_1 \neq k_2$. We choose continuous functions $h_1, h_2 : K \to [0,1]$ such that $h_1(k_1) = 1 = h_2(k_2)$ and $\text{supp}\, h_1 \cap \text{supp}\, h_2 = \emptyset$ ($\text{supp}\, h := \{k \mid k \in K, h(k) \neq 0\}^-$). We have $p(h_2 x_2) \neq 0$ so that $p(h_1 x_1 + b h_2 x_2) = 0$ for a suitable $b \in \mathbb{K}$. Since

$\operatorname{supp} h_1 \cap \operatorname{supp} h_2 = \emptyset$ there is a continuous function $h: K \to [0,1]$ such that $hh_1 = h_1$, $hh_2 = 0$. (*) implies that $p(h_1 x_1) = p(h(h_1 x_1 + h_2 x_2)) = 0$ in contradiction to the choice of k_1.

γ) Suppose that x vanishes in a neighbourhood of k. Choose a continuous function $h: K \to [0,1]$ such that $h(k) = 1$, $hx = 0$. Thus $p(hx) = 0$ and (by the definition of k) consequently $p(x) = 0$.

Now suppose that $x(k) = 0$ but that x does not necessarily vanish on a neighbourhood of k. For arbitrary $\varepsilon > 0$ there is a closed neighbourhood U of k such that $\|x|_U\| \leq \varepsilon$. By lemma 4.2 we may choose $y \in X$ such that $y|_U = x|_U$ and $\|y\| \leq \varepsilon$. $x - y$ vanishes on U so that, by the first part of the proof, $|p(x)| = |p(y)| \leq \|p\|\varepsilon = \varepsilon$. Since ε was arbitrary it follows that $p(x) = 0$.

δ) p_k is defined on all of X_k by 4.1(iii) and γ) implies that p_k is well-defined. It is obvious that p_k is linear and that p_k is continuous with $\|p_k\| \leq \|p\| \leq 1$ (this follows at once from lemma 4.2). $p(x) = p_k(x(k))$ (all $x \in X$) implies that $X_k \neq \{0\}$, and it remains to show that p_k is extreme in the unit ball of $(X_k)'$.

Suppose that $p_k = \frac{1}{2}(p_k^1 + p_k^2)$, where $p_k^1, p_k^2 \in (X_k)'$, $\|p_k^1\|, \|p_k^2\| \leq 1$. Then $p = \frac{1}{2}(p^1 + p^2)$, where $p^i(x) := p_k^i(x(k))$ for $i = 1,2$. Since p is an extreme functional and $\|p^1\|, \|p^2\| \leq 1$ it follows that $p = p^1 = p^2$ so that $p_k = p_k^1 = p_k^2$.

2. Let $p_k \in E_{X_k}$ ($X_k \neq \{0\}$) be given. $p: X \to \mathbb{K}$, $p(x) := p_k(x(k))$ is linear and continuous with $\|p\| \leq \|p_k\|$ and we have to show that p is an extreme functional.

Suppose that $p = \frac{1}{2}(p^1 + p^2)$, where $p^1, p^2 \in X'$, $\|p^1\|, \|p^2\| \leq 1$. We claim that $p^1(x) = p^2(x) = 0$ whenever $x(k) = 0$. Since every element $x \in X$ such that $x(k) = 0$ can be approximated by elements which vanish in a neighbourhood of k (see the proof of γ) above) it suffices to consider the case when $x|_U = 0$ for a suitable neighbourhood U of k.

We have $\|p_k\| = 1$ so that, for arbitrary $\varepsilon > 0$, there is an $\tilde{x}(k) \in X_k$

such that $|p_k(\tilde{x}(k))| \geq 1-\varepsilon$ and $\|\tilde{x}(k)\| = 1$. Replacing, if necessary, $\tilde{x}$ by $h\tilde{x}$ for a suitable $h \in CK$ such that $h(k) = 1$ we may assume that $\|\tilde{x}\| = 1$ and $\tilde{x}|_{K \setminus U} = 0$ (cf. lemma 4.2). It follows that $\|x\| =$ $\| bx + \|x\|\tilde{x}\|$ for every $b \in \mathbb{K}$ such that $|b| = 1$. Accordingly $|p^i(bx + \|x\|\tilde{x})| \leq \|x\|$ for these b and $i = 1,2$.

$$(1-\varepsilon)\|x\| \leq |p(bx + \|x\|\tilde{x})|$$
$$= |\tfrac{1}{2}[p^1(bx + \|x\|\tilde{x}) + p^2(bx + \|x\|\tilde{x})]|$$

implies that $|p^1(bx + \|x\|\tilde{x})|, |p^2(bx + \|x\|\tilde{x})| \geq (1-2\varepsilon)\|x\|$ (all $b \in \mathbb{K}$ such that $|b| = 1$) so that $|p^1(x)|, |p^2(x)| \leq 2\varepsilon\|x\|$ (note that $|p^1(\tilde{x})|, |p^2(\tilde{x})| \leq 1$). This proves that $p^1(x) = p^2(x) = 0$.

It is therefore possible to define $p_k^1, p_k^2 : X_k \to \mathbb{K}$ by $p_k^i(x(k)) := p^i(x)$ $(i=1,2)$. p_k^1, p_k^2 are functionals in the unit ball of $(X_k)'$ such that $p_k = \frac{1}{2}(p_k^1 + p_k^2)$. This yields $p_k = p_k^1 = p_k^2$ so that $p = p^1 = p^2$. □

4.6 Corollary: If X_o is a Banach space and L a locally compact Hausdorff space, then the extreme functionals on $C_o(L,X_o)$ are precisely the mappings $f \mapsto p(f(k))$, where $k \in L$ and $p \in E_{X_o}$. □

Suppose that K and the $(X_k)_{k\in K}$ are as in 4.1, that X is a closed subspace of $\prod\limits_{k\in K}^{\infty} X_k$ and that T_k is an operator on X_k for every $k \in K$. We define $\prod T_k : \prod\limits_{k\in K} X_k \to \prod\limits_{k\in K} X_k$ by $[(\prod T_k)(x)](k) := T_k(x(k))$. If $(\prod T_k)(X) \subset X$, then $\prod T_k$ is continuous on X since this mapping is linear and has a closed graph (this is obvious). If, in addition, X satisfies (i,ii) of 4.1, it follows easily from lemma 4.2 that $\|T_k\| \leq \|\prod T_k\|$ for every $k \in K$.

We will show that for function modules X the operators in the centralizer are of this particularly simple form.

4.7 Proposition: Let $(K,(X_k)_{k\in K},X)$ be a function module and $T:X \to X$ an operator. Then

(i) the multiplication operator $M_h: X \to X$, $x \mapsto hx$, lies in the centralizer of X for every $h \in CK$

(note that M_h is well-defined by def. 4.1(i))

(ii) $T = \Pi T_k$ for a suitable family $(T_k)_{k\in K}$ of operators iff T commutes with $\{ M_h \mid h \in CK\}$

(for simplicity we will write ΠT_k instead of $\Pi T_k|_X$)

(iii) $T \in \mathrm{Mult}(X)$ iff $T = \Pi T_k$, where $T_k \in \mathrm{Mult}(X_k)$ for every $k\in K$

(iv) $T \in Z(X)$ iff $T = \Pi T_k$, where $T_k \in Z(X_k)$ for every $k \in K$ and $(\Pi T_k^*)(X) \subset X$

<u>Proof</u>:

(i) It is easy to see that M_h is M-bounded (M_h satisfies the condition of def. 3.2 with $\lambda = \|h\|$). If h is real-valued it follows that $M_h \in Z(X_{\mathbb{R}})$. In the complex case we write $h = \mathrm{Re}\, h + i\mathrm{Im}\, h$ so that $M_h = M_{\mathrm{Re}\, h} + iM_{\mathrm{Im}\, h} \in Z(X_{\mathbb{R}}) + iZ(X_{\mathbb{R}}) = Z(X)$.

(Alternative proof: by theorem 4.5, M_h is a multiplier for which $M_{\overline{h}}$ is an adjoint so that $M_h \in Z(X)$).

(ii) It is clear that $(\Pi T_k)M_h = M_h(\Pi T_k)$ for every family of operators such that $(\Pi T_k)X \subset X$.

Conversely, let $T: X \to X$ be an operator such that $M_h T = TM_h$ for every $h \in CK$. We claim that $(Tx)(k) = 0$ whenever $x(k) = 0$ $(x \in X, k \in K)$. If $x(k) = 0$ and $\varepsilon > 0$ we choose a neighbourhood U of k such that $\|x(l)\| \leq \varepsilon$ for $l \in U$ and a function $h \in CK$ such that $\|h\| = h(k) = 1$ and $h|_{K \setminus U} = 0$. Accordingly

$$\begin{aligned} \|(Tx)(k)\| &= \|h(k)(Tx)(k)\| \\ &\leq \|(M_h T)(x)\| \\ &= \|TM_h(x)\| \\ &\leq \|T\|\|hx\| \\ &\leq \varepsilon \|T\| \end{aligned}$$

so that $Tx = 0$. As in the proof of th. 4.5 it follows that $T_k: X_k \to X_k$, $x(k) \mapsto (Tx)(k)$ is a well-defined operator with $\|T_k\| \leq \|T\|$. It is obvious that $T = \Pi T_k$.

(iii) This follows from (ii), theorem 4.5, and from the fact that

$\{M_h \mid h \in CK\} \subset Z(X) \subset \mathrm{Mult}(X)$ (recall that Mult(X) is commutative).

(iv) Let T be an operator in Z(X). Since T^* is also a multiplier, we have $T = \Pi T_k$ and $T^* = \Pi S_k$ for suitable families of multipliers $(T_k)_{k\in K}$, $(S_k)_{k\in K}$. It follows from th. 4.5 that $S_k = T_k^*$ for every k. The converse implication is obvious (note that ΠT_k^* is adjoint to ΠT_k by th. 4.5). □

For M-ideals and M-summands we have a similar localization result (prop. 4.9). We prepare the proof by showing that for function modules distances to subspaces can be determined locally.

<u>4.8 Proposition</u>: Let K, $(X_k)_{k\in K}$ be as in 4.1, X, Y subspaces of $\prod_{k\in K} X_k$ such that $Y \subset X$, Y satisfies 4.1(i), and X satisfies 4.1(ii). We define $Y_k := \{y(k) \mid y \in Y\}$ for every $k \in K$. Then

(i) if $x_1,\dots,x_n \in X$ and $r_1,\dots,r_n > 0$, then $\bigcap_{i=1}^n Y \cap B(x_i,r_i) \neq \emptyset$ iff $\bigcap_{i=1}^n Y_k \cap B(x_i(k),r_i) \neq \emptyset$ for every $k \in K$

(ii) $d(x,Y) = \sup_{k\in K} d(x(k),Y_k)$ for every $x \in X$ ($d(x,Y)$ denotes the distance from x to Y)

(iii) $x \in Y^-$ iff $x(k) \in (Y_k)^-$ for every $k \in K$ $(x \in X)$

(iv) Y is dense in X iff $\{x(k) \mid x \in X\} \subset (Y_k)^-$ for every $k \in K$

<u>Proof</u>:

(i) Suppose that $\bigcap_{i=1}^n Y_k \cap B(x_i(k),r_i) \neq \emptyset$ for every $k \in K$. We choose $y^k \in Y$ for $k \in K$ such that $\|(y^k - x_i)(k)\| < r_i$ for $i=1,\dots,n$. Let U_k be a neighbourhood of k such that $\|(y^k - x_i)(l)\| < r_i$ for $l \in U_k$ and $i=1,\dots,n$. K is compact so that there are $k_1,\dots,k_m \in K$ such that $U_{k_1} \cup \dots \cup U_{k_m} = K$. We define $y := \sum_{j=1}^m h_j y^{k_j}$, where $h_1,\dots,h_m$ is a partition of unity subordinate to $U_{k_1},\dots,U_{k_m}$ (i.e. $h_j \in CK$, $0 \leq h_j \leq 1$, $h_j|_{K \setminus U_j} = 0$ for $j=1,\dots,m$, $\sum_{j=1}^m h_j = 1$).

It is easy to see that $\|y - x_i\| < r_i$ for $i=1,\dots,n$ (note that upper semicontinuous functions on compact spaces attain their supremum). We have $y \in Y$ so that $\bigcap_{i=1}^n Y \cap B(x_i,r_i) \neq \emptyset$.

The reverse implication is obvious.

(ii) It is clear that $d(x,Y) \geq \sup_{k \in K} d(x(k),Y_k) =: d$.

Let $\varepsilon > 0$ be arbitrary. We have $Y_k \cap B(x(k),d+\varepsilon) \neq \emptyset$ for every k by definition so that, by (i), $Y \cap B(x,d+\varepsilon) \neq \emptyset$. This proves that $d(x,Y) \leq d+\varepsilon$ and consequently $d(x,Y) \leq d$.

(iii) and (iv) are easy consequences of (ii) since $x \in Y^-$ iff $d(x,Y) = 0$. □

4.9 Proposition: Let X be a function module in $\prod_{k \in K}^{\infty} X_k$, J a closed subspace of X, $J_k := \{x(k) \mid x \in J\}$ for $k \in K$. Then

(i) J is a CK-module (i.e. $M_h J \subset J$ for every $h \in CK$) iff $J = X \cap (\prod J_k)$ $(=\{x \mid x \in X,\ x(k) \in J_k \text{ for every } k \in K\})$

(ii) J is an M-ideal iff J is a CK-module and the J_k are M-ideals in X_k (all $k \in K$);

it follows that every M-ideal is of the form $X \cap (\prod J_k)$, where the J_k are M-ideals in X_k for every $k \in K$

(iii) if J is an M-summand, then the J_k are also M-summands (the converse is not true)

Proof:

(i) It is obvious that $X \cap (\prod J_k)$ is a CK-module. Conversely, suppose that $M_h J \subset J$ for every $h \subset CK$. We have $J \subset X \cap \prod J_k$ by definition, and "$\supset$" follows immediately from prop. 4.8(iii).

(ii) 1. Let J be an M-ideal. J is a CK-module by prop. 4.7(i) and prop. 3.14 so that, by (i), $J = X \cap (\prod J_k)$. We will show that

α) the J_k are closed

β) the J_k have the 3-ball property for open balls

α) Let $(x_n(k))_{n \in \mathbb{N}}$ be a sequence in J_k such that $\sum_{n=1}^{\infty} \|x_n(k)\| < \infty$ $(x_n \in J$ for every $n \in \mathbb{N})$. Choose functions $h_n \in CK$, $h_n(k) = 1$, $\|h_n x_n\| = \|x_n(k)\|$ (lemma 4.2). The $h_n x_n$ are contained in J and $\sum_{n=1}^{\infty} \|h_n x_n\| < \infty$ so that $x := \sum_{n=1}^{\infty} h_n x_n$ exists in J. It is clear that $x(k) = \sum_{n=1}^{\infty} x_n(k) \in J_k$ so that J_k is closed.

β) Let $k \in K$ and $B_i := B(x_i(k), r_i)$ $(i=1,2,3)$ be three open balls such that there are $x(k) \in B_1 \cap B_2 \cap B_3$, $y_i(k) \in J_k \cap B_i$ for $i=1,2,3$ $(x, x_1, x_2, x_3 \in X,\ y_1, y_2, y_3 \in J)$. By lemma 4.2 there are continuous functions $h_1, h_2, h_3, h_1', h_2', h_3' : K \to [0,1]$ such that $h_i(k) = h_i'(k) = 1$, $\| h_i(x-x_i) \| = \| (x-x_i)(k) \| < r_i$, $\| h_i'(x_i-y_i) \| = \| (x_i-y_i)(k) \| < r_i$ $(i=1,2,3)$. With $h := \inf\{h_1, h_2, h_3, h_1', h_2', h_3'\}$ it follows that $\| h(x-x_i) \|, \| h(y_i-x_i) \| < r_i$ so that (since J is a CK-module), $J \cap B(hx_i, r_i) \neq \emptyset$ $(i=1,2,3)$ and $\bigcap_{i=1}^{3} B(hx_i, r_i) \neq \emptyset$. Since J is an M-ideal there is a $y \in J$ such that $y \in J \cap \bigcap_{i=1}^{3} B(hx_i, r_i)$. Accordingly $y(k) \in J_k \cap \bigcap_{i=1}^{3} B(x_i(k), r_i)$ so that J_k is an M-ideal.

2. Suppose that $J = X \cap (\prod J_k)$ where $J_k = \{ x(k) \mid x \in J \}$ is an M-ideal in X_k for every $k \in K$. Every J_k has the 3-ball property for open balls, and proposition 4.8(i) implies that this is also true for J. Thus J is an M-ideal.

(iii) Let J be an M-summand of X and $k \in K$. We define $J_k := \{x(k) \mid x \in J\}$ and $J_k^{\perp} := \{x(k) \mid x \in J^{\perp}\}$ and we claim that $J_k^{\perp}$ and J_k are complementary M-summands. $J + J^{\perp} = X$ implies that $J_k + J_k^{\perp} = X_k$. For $x \in J$ and $x^{\perp} \in J^{\perp}$ we choose continuous functions $h_1, h_2, h_3 : K \to [0,1]$ such that $h_1(k) = h_2(k) = h_3(k) = 1$ and $\| h_1(x+x^{\perp}) \| = \| (x+x^{\perp})(k) \|$, $\| h_2 x \| = \| x(k) \|$, $\| h_3 x^{\perp} \| = \| x^{\perp}(k) \|$ (see lemma 4.2). Then, with $h := \inf\{ h_1, h_2, h_3 \}$ it follows that $hx \in J$ and $hx^{\perp} \in J^{\perp}$ so that

$$
\begin{aligned}
\| (x + x^{\perp})(k) \| &= \| h(x + x^{\perp}) \| \\
&= \max\{ \| hx \|, \| hx^{\perp} \| \} \\
&= \max\{ \| x(k) \|, \| x^{\perp}(k) \| \}.
\end{aligned}
$$

Thus J_k and $J_k^{\perp}$ satisfy the conditions of lemma 1.2(ii) so that they are complementary M-summands.

The converse is not true: if X is the function module CK in $\prod_{k \in K}^{\infty} \mathbb{K}_k$ (K a compact Hausdorff space, $\mathbb{K}_k := \mathbb{K}$ for every $k \in K$) and $L \subset K$ closed, then $J_L := \{ f \mid f \in CK,\ f|_L = 0 \}$ is an M-ideal but in general not an M-summand. However, the $(J_L)_k$ are M-summands for every $k \in K$ $\square$

4.10 Corollary: Let $(K,(X_k)_{k\in K},X)$ be a function module.

(i) $J_L := \{x \mid x\in X,\ x|_L = 0\}$ is an M-ideal for every closed subset L of K

(ii) If $L\subset K$ is clopen, then J_L is an M-summand of X

Proof:

(i) J_L is a CK-module, and $(J_L)_k = \{0\}$ if $k\in L$ (this is trivial) and $(J_L)_k = X_k$ if $k\notin L$ (this follows from 4.1(i)(iii)). Thus J_L is an M-ideal by prop. 4.9(ii).

(ii) Let h be the characteristic function of $K\setminus L$. It is obvious that M_h is an M-projection with range J_L. □

Multiplication operators on function modules play an important role. We proved in prop. 4.7(i) that the M_h with $h\in CK$ define operators in the centralizer. The following proposition shows that this is also true for a class of functions on K which in general is strictly larger than CK.

4.11 Proposition: Let $(K,(X_k)_{k\in K},X)$ be a function module and $\alpha: K\to \mathbb{K}$ a (not necessarily continuous) function such that $M_\alpha X\subset X$ ($(M_\alpha x)(k) := \alpha(k)x(k)$). Then

(i) α is bounded on $\{k \mid k\in K,\ X_k \neq \{0\}\}$

(ii) M_α is contained in the strong operator closure of $\{M_h \mid h\in CK\}$

(iii) $M_\alpha \in Z(X)$; the adjoint operator is just the operator $M_{\overline{\alpha}}$

(iv) let $\mathfrak{T}_{ex}$ be the coarsest topology on B(X) for which the functionals $T\mapsto p(Tx)$ (all $p\in E_X$, $x\in X$) are continuous; then the closure of $\{M_h \mid h\in CK\}$ with respect to this topology is contained in Z(X); all operators in this closure are of the form M_β, where $\beta: K\to\mathbb{K}$ is a function such that $M_\beta X\subset X$

(v) the closures in B(X) of $\{M_h \mid h\in CK\}$ with respect to the weak operator topology and the strong operator topology coincide; these closures are just $\{M_\beta \mid \beta: K\to\mathbb{K},\ M_\beta X\subset X\}$

(vi) suppose that $Z(X_k) = \mathbb{K}\,Id$ for every $k \in K$; then $Z(X) =$

$\{M_\alpha \mid \alpha : K \to \mathbb{K},\ M_\alpha X \subset X\}$

<u>Proof</u>:

(i) M_α is just the operator $\prod_{k\in K} (\alpha(k) Id)$ (cf. the remarks before prop. 4.7). Therefore M_α is continuous, and $\|\alpha(k)Id\| \le \|M_\alpha\|$ for every $k \in K$.

This yields $\sup\{|\alpha(k)| \mid X_k \ne \{0\}\} = \sup\{\|\alpha(k)Id\| \mid X_k \ne \{0\}\} \le \|M_\alpha\|$.

(ii) Let $x_1,\ldots,x_n$ be a finite family in X and $\varepsilon > 0$. We have to show that there is a function $h \in CK$ such that $\|M_h x_i - M_\alpha x_i\| = \|(h - \alpha)x_i\| \le \varepsilon$ $(i=1,\ldots,n)$.

Let $\tilde{X} := \{(\tilde{x}_1,\ldots,\tilde{x}_n) \mid \tilde{x}_1,\ldots,\tilde{x}_n \in X\} \subset \prod^\infty_{k\in K} (X_k)^n$ ($(X_k)^n$ is the product of n copies of X_k, provided with the supremum norm) and $\tilde{Y} := \{(hx_1,\ldots,hx_n) \mid h \in CK\} \subset \tilde{X}$. $\tilde{X}$ and $\tilde{Y}$ satisfy the conditions of prop. 4.8 and $d((\alpha x_1,\ldots,\alpha x_n)(k),\tilde{Y}_k) = 0$ for every $k \in K$ (since $(\alpha x_1,\ldots,\alpha x_n)(k) = (\alpha(k)x_1,\ldots,\alpha(k)x_n)(k) \in \tilde{Y}_k$) so that, by prop. 4.8(ii), $d((\alpha x_1,\ldots,\alpha x_n),\tilde{Y}) = 0$. Thus there is a function $h \in CK$ such that $\|(\alpha x_1,\ldots,\alpha x_n) - (hx_1,\ldots,hx_n)\| = \max_i \|\alpha x_i - hx_i\| \le \varepsilon$.

(iii) Let $x \in X$ and $\varepsilon > 0$. By (ii) there is a function $h \in CK$ such that $\|(\alpha - h)x\| \le \varepsilon$. Thus $\|(\overline{\alpha} - \overline{h})x\| \le \varepsilon$ so that, since $\overline{h}x \in X$, $d(\overline{\alpha}x, X) \le \varepsilon$. This proves that $\overline{\alpha}x \in X^- = X$.

It is clear from theorem 4.5 that M_α is a multiplier and that $(M_\alpha)^* = M_{\overline{\alpha}}$.

(iv) Suppose that T is an operator in this closure. By (iii) it suffices to show that $T = M_\beta$ for a suitable function $\beta : K \to \mathbb{K}$.

For $k \in K$ such that $X_k \ne \{0\}$, $x_1, x_2 \in X$, $p_1, p_2 \in E_{X_k}$ $(\subset E_X)$ and $\varepsilon > 0$ there is a function $h \in CK$ such that $|p_i((Tx_j - hx_j)(k))| = |p_i((Tx_j - h(k)x_j)(k))| \le \varepsilon$ $(i,j = 1,2)$. With $\varepsilon \to 0$ we obtain a $\beta_{x_1x_2p_1p_2} \in \mathbb{K}$ such that $p_i((Tx_j)(k)) = \beta_{x_1x_2p_1p_2} p_i(x_j(k))$ $(i,j=1,2)$. Since this is true for all pairs x_1, x_2 and p_1, p_2 it follows that there is a common value, say β_k, such that $p((Tx)(k)) = \beta_k p(x(k))$ for $p \in E_{X_k}$ and $x \in X$. Thus $Tx(k) = \beta_k x(k)$ for every $x \in X$, i.e. $T = M_\beta$

(where $\beta(k) := \beta_k$ if $X_k \neq \{0\}$ and $\beta(k) := 0$ if $X_k = \{0\}$).

(v) This follows from (ii) and (iv) (note that $\mathfrak{T}_{ex}$ is coarser than the weak operator topology).

(vi) This is a consequence of (iii) and proposition 4.7(iv) □

B. Function module representations

In this section we will prove that every Banach space can be represented as a function module in such a way that the base space is "as large as possible".

4.12 Definition: Let X be a Banach space. A function module representation $[\rho,(K,(X_k)_{k\in K},\tilde{X})]$ of X is a function module $(K,(X_k)_{k\in K},\tilde{X})$ together with an isometric isomorphism $\rho: X \to \tilde{X}$.

4.13 Lemma: Suppose that $[\rho,(K,(X_k)_{k\in K},\tilde{X})]$ is a function module representation of the Banach space X. Then

(i) $\rho^{-1}M_h\rho \in Z(X)$ for every $h \in CK$

(ii) $h \mapsto \rho^{-1}M_h\rho$ is an isometric isomorphism of B^*-algebras from CK onto $Z_\rho(X) := \{\rho^{-1}M_h\rho \mid h \in CK\} \subset Z(X)$

Proof:

(i) $(\rho^{-1})'$ is an isometric isomorphism from X' onto $\tilde{X}'$ so that $(\rho^{-1})'(E_X) = E_{\tilde{X}}$. For $p \in E_X$ we have $((\rho^{-1})'(p))\circ M_h = a_{M_h}((\rho^{-1})'(p))((\rho^{-1})'(p))$ since M_h is a multiplier so that $p\circ(\rho^{-1}\circ M_h\circ\rho) = a_{M_h}(p\circ\rho^{-1})p$, i.e. $\rho^{-1}\circ M_h\circ\rho$ is a multiplier such that $a_{(\rho^{-1}M_h\rho)}(p) = a_{M_h}(p\circ\rho^{-1})$. Since $a_{(\rho^{-1}M_{\bar{h}}\rho)} = \overline{a_{(\rho^{-1}M_h\rho)}}$ it follows that $\rho^{-1}M_{\bar{h}}\rho$ is adjoint to $\rho^{-1}M_h\rho$, i.e. $\rho^{-1}M_h\rho$ lies in the centralizer.

(ii) It follows from (i) that $h \mapsto \rho^{-1}M_h\rho$ is a well-defined homomorphism of B^*-algebras. We have $\|\rho^{-1}M_h\rho\| = \|M_h\|$ since ρ is an isometric isomorphism, and $\|M_h\| = \|h\|$ is an easy consequence of 4.1(i)(iv) and lemma 4.2. This proves that $h \mapsto \rho^{-1}M_h\rho$ is isometric □

Thus with every function module representation of X there is associated a class of operators in Z(X) which are, if we identify X with $\tilde{X}$, multiplication operators.

For example:

- if $[\rho,(\{1\},X,X)]$ is the trivial representation of X of example 3 on p. 78, then $Z_\rho(X) = \mathbb{K}\,Id$
- let c_o be represented as a function module with base space $\alpha\mathbb{N}$, $X_k = \mathbb{K}$ for $k \in \mathbb{N}$ and $X_\infty := \{0\}$; then $Z_\rho(c_o)$ corresponds to the subalgebra c of $m \cong Z(c_o)$

The main theorem of this section states that there is a function module representation of every Banach space X such that all operators in Z(X) are multiplication operators:-

4.14 Theorem: Let X be a Banach space and K_X as in prop. 3.10. There exists a function module representation $[\rho,(K_X,(X_k)_{k\in K_X},\tilde{X})]$ of X such that $Z_\rho(X) = Z(X)$.

Thus every Banach space can be regarded as a function module in a suitable product $\prod\limits_{k\in K_X}^{\infty} X_k$ such that the operators in the centralizer of X are precisely the operators M_h, $h \in CK$.

Proof: By prop. 3.10 there exists an isometric isomorphism ν of B^*-algebras from Z(X) onto $C(K_X)$. The composition of ν^{-1} with the mapping $T \mapsto a_T(p)$ (from Z(X) to $\mathbb{K}$; $p \in E_X$ a fixed functional) is therefore a homomorphism of B^*-algebras and thus, by prop. 0.2, the evaluation functional associated with a point k(p) of K_X:

$$(*)\quad \begin{array}{l} \text{for } p \in E_X \text{ there exists a } k(p) \in K_X \text{ such that } a_T(p) = \nu(T)(k(p)) \\ \text{for every } T \in Z(X) \end{array}$$

Now assume that T_1, T_2 are operators in the centralizer such that $|(\nu(T_1))(k)| \leq |(\nu(T_2))(k)|$ for every $k \in K$ and that $x \in X$. We choose $p \in E_X$ such that $|p(T_1x)| = \|T_1x\|$, and it follows from (*)

that
$$\begin{aligned}\|T_1x\| &= |p(T_1x)| \\ &= |a_{T_1}(p)|\,|p(x)| \\ &= |\nu(T_1)(k(p))|\,|p(x)| \\ (**) \qquad &\leq |\nu(T_2)(k(p))|\,|p(x)| \\ &= |a_{T_2}(p)|\,|p(x)| \\ &= |p(T_2x)| \\ &\leq \|T_2x\| .\end{aligned}$$

We are now going to define a function module representation of X with base space K_X.

[To motivate the following construction assume for the moment that X _is_ already a function module with base space K_X. The problem is to reconstruct the X_k from K_X and the Banach space X. Since X_k is, up to isometric isomorphism, the quotient $X/\{x \mid x \in X, \|x(k)\| = 0\}$ (this follows from 4.1(iii) and lemma 4.2) it is sufficient to reconstruct the $\|x(k)\|$ for $x \in X$ and $k \in K_X$. However, these numbers can easily be determined since $\|x(k)\| = \inf\{\|hx\| \mid h \in CK,\ h \geq 0,\ h(k) > 1\}$ as a consequence of the upper semicontinuity of the $\|x(\cdot)\|$.]

For $x \in X$ let $|x| : K_X \to \mathbb{R}$ (the <u>norm resolution</u> of x) be the function $|x|(k) := \inf\{\|Tx\| \mid T \in W_k\}$, where $W_k :=$ $\{T \mid T \in Z(X),\ \nu(T) \geq 0,\ (\nu(T))(k) > 1\}$. We claim that

α) $x \mapsto |x|(k)$ is a seminorm for every $k \in K$

β) $|x|$ is upper semicontinuous for every x

γ) $|Tx - (\nu(T))(k)x|(k) = 0$ and consequently $|Tx|(k) = |(\nu(T))(k)|\,|x|(k)$ for $T \in Z(X)$, $x \in X$, $k \in K_X$

δ) if X_k is the completion of the normed space associated with the semi-normed space $(X, |\cdot|(k))$ (i.e. the completion of $X/\{x \mid x \in X, |x|(k)=0\}$, provided with the norm $[x]_k \mapsto |x|(k)$) and $\rho : X \to \prod^{\infty}_{k \in K_X} X_k$ the mapping $x \mapsto (k \mapsto [x]_k)$, then $[\rho, (K_X, (X_k)_{k \in K_X}, \rho(X))]$ is a function module representation of X

$\varepsilon)\ Z_\rho(X) = Z(X)$.

α) For $x,y \in X$ and $T_1, T_2 \in W_k$ we choose a $T \in W_k$ such that $\nu(T) \leq \inf\{\nu(T_1), \nu(T_2)\}$ (for example, $T := \nu^{-1}(\inf\{\nu(T_1), \nu(T_2)\})$. It follows from (**) that $\|T(x+y)\| \leq \|Tx\| + \|Ty\| \leq \|T_1x\| + \|T_2y\|$ so that $|x+y|(k) \leq |x|(k) + |y|(k)$.

It is obvious that $|x|(k) \geq 0$ and $|\lambda x|(k) = |\lambda|\,|x|(k)$ for $x \in X$ and $\lambda \in \mathbb{K}$.

β) Suppose that $|x|(k) < r$ ($k \in K_X$ and $x \in X$ fixed but arbitrary). We choose an operator $T \in W_k$ such that $\|Tx\| < r$. Since $\nu(T)$ is continuous at k, we have $T \in W_l$ for l in a suitable neighbourhood U of k. Thus $|x|(l) \leq \|Tx\| < r$ for these l, i.e. $|x|$ is upper semicontinuous at k.

γ) For $k \in K_X$ and $T \in Z(X)$ the function $\nu(T) - \nu(\nu(T)(k)\,\mathrm{Id})$ vanishes at k. Thus, for $\varepsilon > 0$, there is a $T_\varepsilon \in W_k$ such that $\|\nu(T_\varepsilon)\,[\nu(T) - \nu(T)(k)]\| \leq \varepsilon$. Accordingly $|Tx - \nu(T)(k)x|(k) \leq \|T_\varepsilon(T - \nu(T)(k)\mathrm{Id})x\| \leq \varepsilon\|x\|$ so that $|Tx - \nu(T)(k)x|(k) = 0$.

δ) ρ is an isometry since we may choose $p \in E_X$ for $x \in X$ with $|p(x)| = \|x\|$ so that (by (*) and the definition of $W_{k(p)}$)

$$
\begin{aligned}
\|\rho(x)\| &\geq \|\rho(x)(k(p))\| \\
&= |x|(k(p)) \\
&= \inf\{\|Tx\| \mid T \in W_{k(p)}\} \\
&\geq \inf\{|p(Tx)| \mid T \in W_{k(p)}\} \\
&= \inf\{|a_T(p)p(x)| \mid T \in W_{k(p)}\} \\
&= \inf\{|(\nu(T))(k(p))|\,|p(x)| \mid p \in W_{k(p)}\} \\
&\geq |p(x)| \\
&= \|x\| .
\end{aligned}
$$

$\|\rho(x)\| \leq \|x\|$ follows from $|x|(k) \leq \inf\{\|(1+\varepsilon)\mathrm{Id}(x)\| \mid \varepsilon > 0\} = \|x\|$ (all $x \in X$).

$\rho(X)$ satisfies 4.1(ii) by β) and 4.1(i) is valid since $\nu(T)\rho(x) =$

$\rho(Tx)$ as a consequence of γ). (iii)' of cor. 4.4 is satisfied by definition so that, by cor. 4.4, it remains to show that $\{ k \mid k \in K_X, X_k \neq \{0\}\}$ is dense in K_X.
For arbitrary $\nu(T) \in C(K_X) \setminus \{0\}$ we choose $x \in X$ such that $Tx \neq 0$. Since ρ is injective, there must be a point $k \in K_X$ such that $(\rho(Tx))(k) = (\nu(T))(k)(\rho(x))(k) \neq 0$. It follows that every non-zero continuous function on K_X is different from zero at a suitable point of $\{k \mid k \in K_X, X_k \neq \{0\}\}$ so that this set must be dense in K_X.

ε) For $T \in Z(X)$ and $x \in X$ we have $\rho(Tx) = \nu(T)\rho(x)$ by γ) so that $\rho \circ T = M_{\nu(T)} \circ \rho$ and consequently $T = \rho^{-1} M_{\nu(T)} \rho$. Hence

$$
\begin{aligned}
Z_\rho(X) &= \{ \rho^{-1} M_h \rho \mid h \in CK_X\} \\
&= \{ \rho^{-1} M_{\nu(T)} \rho \mid T \in Z(X) \} \\
&= Z(X).
\end{aligned}
$$
□

We are now going to show that the function module representation in th. 4.14 is essentially unique and, in a sense, the "finest" representation of X as a function module.

<u>4.15 Definition</u>: Let X be a Banach space and $R_i := [\rho_i, (K_i, (X^i_k)_{k \in K_i}, X_i)]$ (i=1,2) function module representations of X. (i) We say that R_1 is <u>finer</u> than R_2 (and we will write $R_2 \lesssim R_1$ in this case) if there are

- a continuous map t from K_1 onto K_2
- a family of isometric isomorphisms $S_l : X_1|_{t^{-1}(l)} \to X^2_l$ (all $l \in K_2$; for the definition of $X_1|_{t^{-1}(l)}$ see prop. 4.3) such that $S \circ \rho_1 = \rho_2$ (where $(Sx_1)(l) := S_l(x_1|_{t^{-1}(l)})$ for $x_1 \in X_1$ and $l \in K_2$):

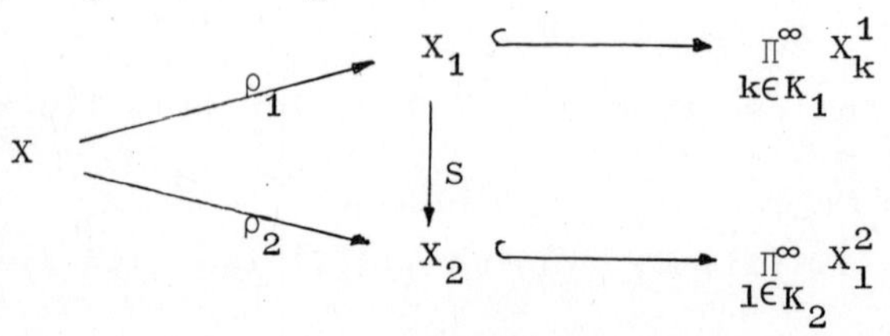

(ii) R_1 and R_2 are said to be <u>equivalent</u> ($R_2 \approx R_1$) if $R_2 \precsim R_1$ and, in addition, the mapping t in (i) is a homeomorphism (so that the S_1 are isometric isomorphisms from $X^1_{t^{-1}(1)}$ onto X^2_1)

<u>Remarks</u>: 1. We will see at once that $R_2 \approx R_1$ iff $R_1 \precsim R_2$ and $R_2 \precsim R_1$.

2. The reader should observe that, if R_1 is finer than R_2, R_1 is obtained from R_2 by "splitting up" the component spaces X^2_1 as subspaces of $\prod^{\infty}_{k\in t^{-1}(1)} X^1_k$ (in fact, apart from 4.1(iv) all conditions of def. 4.1 for function modules are satisfied for X^2_1 in $\prod^{\infty}_{k\in t^{-1}(1)} X^1_k$):

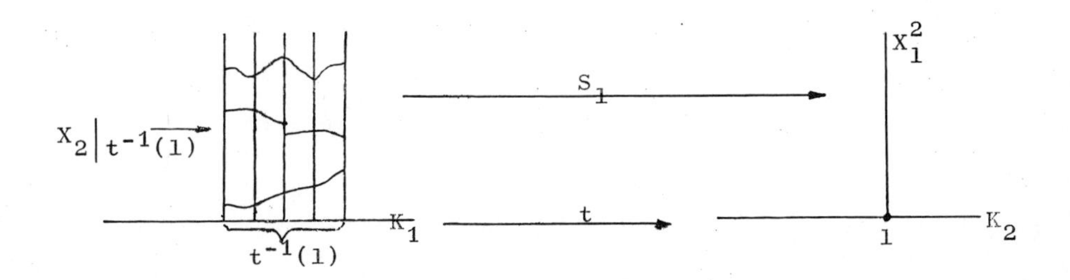

<u>fig. 12</u>

Equivalence of R_1 and R_2 means that $\rho_1(X)$ and $\rho_2(X)$ differ only by a permutation of the base space.

<u>4.16 Theorem</u>: Let $R_i := [\rho_i,(K_i,(X^i_k)_{k\in K_i},X_i)]$ (i=1,2) be two function module representations of the Banach space X.

Then (i) $R_2 \precsim R_1$ iff $Z_{\rho_2}(X) \subset Z_{\rho_1}(X)$

(ii) $R_2 \approx R_1$ iff $Z_{\rho_2}(X) = Z_{\rho_1}(X)$

(which implies that $R_2 \approx R_1$ iff $R_2 \precsim R_1$ and $R_1 \precsim R_2$)

<u>Proof</u>:

<u>(i) "⇒"</u>: Suppose that $R_2 \precsim R_1$ and that t, $(S_l)_{l\in K_2}$ and S are as in def. 4.15(i). For $h \in CK_2$ and $l \in K_2$ the function $h\circ t$ has the con-

stant value h(1) on $t^{-1}(1)$ so that $[(h\circ t)x_1]|_{t^{-1}(1)} = h(1)[x_1|_{t^{-1}(1)}]$ for every $x_1 \in X_1$. Thus $h\cdot(Sx_1) = S((h\circ t)x_1)$ so that $M_h S = SM_{h\circ t}$. $S\circ\rho_1 = \rho_2$ implies that $\rho_2^{-1}M_h\,\rho_2 = \rho_1^{-1}M_{h\circ t}\,\rho_1$, i.e.

$$
\begin{aligned}
Z_{\rho_2}(X) &= \{\rho_2^{-1}M_h\,\rho_2 \mid h \in CK_2\} \\
&\subset \{\rho_1^{-1}M_{\tilde h}\,\rho_1 \mid \tilde h \in CK_1\} \\
&= Z_{\rho_1}(X).
\end{aligned}
$$

"$\Leftarrow$": By lemma 4.13(ii) there exists an isometric isomorphism ω of B*-algebras from CK_2 onto a closed self-adjoint subalgebra of CK_1 (ω satisfies

$$(*) \qquad \rho_2^{-1}M_h\,\rho_2 = \rho_1^{-1}M_{\omega(h)}\,\rho_1 \qquad \text{for all } h \in CK_2 \quad .)$$

Thus there is a continuous map t from K_1 onto K_2 such that $\omega(h) = h\circ t$ for $h \in CK_2$ (prop. 0.2). We have to show that there are isometric isomorphisms $S_l : X_1|_{t^{-1}(l)} \to X_l^1$ such that $S_l(\rho_1(x)|_{t^{-1}(l)}) = \rho_2(x)(l)$ for $l \in L$ and $x \in X$.

Because of this we are motivated to define S_l by

$$S_l(\rho_1(x)|_{t^{-1}(l)}) := \rho_2(x)(l) \;,$$

and we have to prove that

α) S_l is well-defined

β) S_l is an isometric isomorphism.

α) Let $x \in X$ be given such that $\rho_1(x)|_{t^{-1}(l)} = 0$ (we have to show that $\rho_2(x)(l) = 0$ in this case).

For $\varepsilon > 0$ the set $V_\varepsilon := \{k \mid k \in K_1, \|\rho_1(x)(k)\| \geq \varepsilon\}$ is compact in K_1 so that, since V_ε and $t^{-1}(l)$ are disjoint, $t(V_\varepsilon)$ is a compact subset of K_2 which does not contain l. Choose $h \in CK_2$ such that $\|h\| = h(l) = 1$ and $h|_{t(V_\varepsilon)} = 0$. It follows that $\|(h\circ t)\,\rho_1(x)\| \leq \varepsilon$ so that, since ρ_1^{-1} and ρ_2^{-1} are isometric isomorphisms,

$$
\begin{aligned}
\|\rho_2(x)(l)\| &\leq \|(M_h\circ\rho_2)(x)\| \\
&= \|(\rho_2^{-1}M_h\rho_2)(x)\| \\
&= \|(\rho_1^{-1}M_{h\circ t}\,\rho_1)(x)\|
\end{aligned}
$$

$$= \|(h\circ t)\ \rho_1(x)\|$$
$$\leq \varepsilon \quad .$$

This proves that $\rho_2(x)(l) = 0$.

β) $\underline{S_1 \text{ is onto}}$:

This follows from def. 4.1(iii) (recall that $\rho_2(X)$ is a function module in $\prod^{\infty}_{l \in K_2} X_l^2$).

$\underline{S_1 \text{ is an isometry}}$:

For $x \in X$ there are
- a $y \in X$ such that $\rho_1(x)|_{t^{-1}(l)} = \rho_1(y)|_{t^{-1}(l)}$ and $\|y\| = \|\rho_1(y)\| = \|\rho_1(x)|_{t^{-1}(l)}\|$
- a function $h \in CK_2$ such that $h(l) = 1$ and $\|\rho_2(x)(l)\| = \|h\,\rho_2(x)\|$

(this follows from lemma 4.2). Accordingly

$$\begin{aligned}
\| S_1(\rho_1(x)|_{t^{-1}(l)})\| &= \|S_1(\rho_1(y)|_{t^{-1}(l)})\| \\
&= \|\rho_2(y)(l)\| \\
&\leq \|y\| \\
&= \|\rho_1(x)|_{t^{-1}(l)}\| \\
&\leq \|(h\circ t)\ \rho_1(x)\| \qquad (\text{since } h\circ t|_{t^{-1}(l)} = 1) \\
&= \|(M_{h\circ t}\ \rho_1)(x)\| \\
&= \|(\rho_1^{-1} M_{h\circ t}\ \rho_1)(x)\| \\
&= \|(\rho_2^{-1} M_h\ \rho_2)(x)\| \\
&= \|(M_h\ \rho_2)(x)\| \\
&= \|(\rho_2(x))(l)\| \\
&= \|S_1(\rho_1(x)|_{t^{-1}(l)})\|
\end{aligned}$$

which proves that $\|S_1(\rho_1(x)|_{t^{-1}(l)})\| = \|\rho_1(x)|_{t^{-1}(l)}\|$.

$\underline{\text{(ii)"}\Rightarrow\text{"}}$: It can easily be shown that $R_2 \approx R_1$ implies $R_1 \approx R_2$ and $R_1 \lesssim R_2$ so that the assertion is a consequence of (i).

$\underline{\text{"}\Leftarrow\text{"}}$: The mapping t in the construction of the proof of (i) must be injective in this case since CK_2 and CK_1 are isometrically isomorphic.

□

We restate th. 4.16(ii) for the case of function modules. In view of part II this is one of the most important results of this section:

4.17 Corollary: Suppose that $(K,(X_k)_{k\in K},X)$ and $(L,(Y_l)_{l\in L},Y)$ are function modules such that $Z(X) = \{M_h \mid h \in CK\}$ and $Z(Y) = \{M_h \mid h \in CL\}$. Then for every isometric isomorphism $I:X \to Y$ there are

- a homeomorphism $\tilde{t}:L \to K$
- a family of isometric isomorphisms $S_l:X_{\tilde{t}(l)} \to Y_l$ (all $l\in L$)

such that $(Ix)(l) = S_l(x(\tilde{t}(l)))$ for $x\in X$ and $l \in L$.
In particular, X and Y are isometrically isomorphic only if K and L are homeomorphic and the families $(X_k)_{k\in K}$ and $(Y_l)_{l\in L}$ contain the same spaces (modulo isometric isomorphism)

Proof: Let $\tilde{R}$ be the identical representation of X in $\prod^{\infty}_{k\in K} X_k$ and R the representation $[I,(L,(Y_l)_{l\in L},Y)]$ of X. Since the algebras $Z_\rho(X)$ are just $Z(X)$ for both representations it follows that $R \approx \tilde{R}$. With S as in def. 4.15 we have $S\tilde{\rho} = I$ so that (since $\tilde{\rho} = Id$) $S = I$. The assertion follows with t, $(S_l)_{l\in L}$ as in 4.15 and $\tilde{t} := t^{-1}$. □

4.18 Corollary: Let X be a Banach space. $R \mapsto Z_\rho(X)$ induces a one-to-one correspondence between the function module representations of X modulo equivalence and the closed self-adjoint subalgebras of Z(X) which contain the identity operator. Since, by th. 4.16, this correspondence is compatible with the order structures it follows that the equivalence classes of function module representations of X form a complete lattice.

Proof: Let A be a closed self-adjoint subalgebra of Z(X) which contains the identity operator. We only have to show that there is a function module representation R of X such that $Z_\rho(X) = A$ (the other assertions follow from th. 4.16).
Since $\nu(A)$ (ν as in the proof of th. 4.14) is a closed self-adjoint subalgebra with $\underline{1} \in \nu(A)$ there exist a compact Hausdorff

space L and a continuous mapping t from K_X onto L such that $\nu(A) = \{h \circ t \mid h \in CL\}$ (prop. 0.2). Then, with $[\rho,(K_X,(X_k)_{k\in K_X}, \rho(X))]$ as in th. 4.14 we define $Y_l := \rho(X)|_{t^{-1}(l)}$ for $l \in L$ and $\tilde{\rho} : X \to \prod^{\infty}_{l\in L} Y_l$ by $(\tilde{\rho}(x))(l) := \rho(x)|_{t^{-1}(l)}$. It is routine to show that $[\tilde{\rho},(L,(Y_l)_{l\in L}, \tilde{\rho}(X))]$ is a function module representation of X such that $Z_{\tilde{\rho}}(X) = A$ (note that this is the same construction as in example 6 on p. 78). □

<u>Remark</u>: Since $Z(X) \cong C(K_X)$ and the closed self-adjoint subalgebras of $C(K_X)$ which contain $\underline{1}$ are in one-to-one correspondence with the quotients of K_X (prop. 0.2) we may restate the corrolary by saying that there are as many inequivalent function module representations of X as there are quotients of K_X.
In particular, if X and K are given, we are able to decide whether X can be represented as a function module with base space K: this is possible iff K is a quotient of K_X.

By th. 4.14 and th. 4.16(ii) there is essentially one way to represent a Banach space X as a function module such that the centralizer contains precisely the multiplication operators associated with the continuous scalar-valued functions on the base space. The base space and the componeent spaces are uniquely determined.
Such a representation will be called a <u>maximal function module representation</u>.
We will see in part II that the question whether X has the Banach-Stone property (def. 8.2) or not can often be decided if a maximal function module representation of X is known.

The construction in th. 4.14 is not always the simplest way to get such a representation. Often it is more natural to represent X as a function module $(K,(X_k)_{k\in K},X)$ in such a way that the choice of K and the X_k is motivated by the structure of X. If it is possible to show that $Z(X) = \{M_h \mid h \in CK\}$, then the representation is maximal (so that, in particular, $K \cong K_X$).

Examples:

1. Since $Z(C_oL) = \{M_h \mid h \in C^bL\}$ (see p. 63) it follows that the function module representation of C_oL on p. 78 (where $X_o = \mathbb{K}$ and $K = \beta L$) is maximal (L a nonvoid locally compact Hausdorff space.
2. Let s be a scalar such that $0 < |s| \leq 1$ and $X_s :=$ $\{f \mid f \in C[0,2\pi],\ f(2\pi) = sf(0)\}$. We represent X_s as a function module in $\prod^\infty_{k \in K} \mathbb{K}_k$, where $\mathbb{K}_k := \mathbb{K}$ for every $k \in K :=$ $\{e^{i\Theta} \mid \Theta \in [0,2\pi]\}$.

   For $f \in X_s$ we identify f with the element $e^{i\Theta} \mapsto f(\Theta)$ of $\prod^\infty_{k \in K} \mathbb{K}_k$.

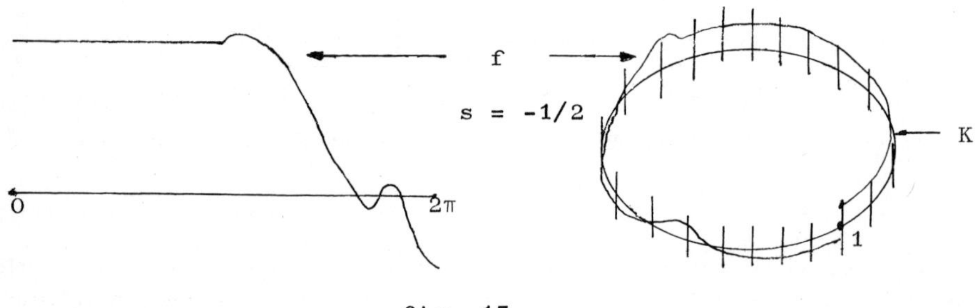

fig. 13

It is routine to prove that this is a function module representation of X_s (it is essential for the upper semicontinuity of the norm resolutions at k = 1 that $|s| \leq 1$).
Since the component spaces are one-dimensional, every operator $T \in Z(X_s)$ is a multiplication operator, say M_α (prop. 4.11(vi)). $M_\alpha X_s \subset X_s$ implies that α is continuous on K so that the representation is maximal.
Note: This example will be of interest for our investigations in chapter 9 and chapter 11 below.

3. Let S_n be the n-dimensional sphere (i.e. the surface of the unit ball in the (n+1)-dimensional Hilbert space) and $X :=$ $\{f \mid f \in C(S_n),\ f(-s) = -f(s) \text{ for every } s \in S_n\}$, X provided with the supremum norm. We claim that $Z(X) =$ $\{M_h \mid h \in C(S_n),\ h(-s) = h(s) \text{ for every } s \in S_n\} =: \tilde{Z}(X)$:

It is clear that $M_h X \subset X$ and that M_h is M-bounded so that $M_h = M_{\mathrm{Re}\, h} + iM_{\mathrm{Im}\, h} \in Z(X_{\mathbb{R}}) + iZ(X_{\mathbb{R}}) = Z(X)$ for $M_h \in \tilde{Z}(X)$.

Conversely, let $T \in Z(X)$ be arbitrary. Since T commutes with $\tilde{Z}(X)$ by the first part of the proof it follows that $f(s) = 0$ implies that $(Tf)(s) = 0$ for $f \in X$ and $s \in S_n$ (sketch of proof: for $\varepsilon > 0$, choose $M_h \in \tilde{Z}(X)$ such that $h(s) = 1$ and $\|M_h f\| \leq \varepsilon$; then $|(Tf)(s)| = |(M_h \cdot (Tf))(s)| \leq \|TM_h f\| \leq \varepsilon \|T\|$).

Thus there is a function $h: S_n \to \mathbb{K}$ such that $Tf = hf$ for every $f \in X$. $TX \subset X$ implies that $T = M_h \in \tilde{Z}(X)$.

Let K_n be the quotient $S_n/\sim$, where $\sim$ is the relation $s_1 \sim s_2$ iff $s_1 = \pm s_2$. Since $CK_n \cong \{h \mid h \in C(S_n), h(-s) = h(s) \text{ for every } s \in S_n\}$ it follows that $K_X = K_n$.

The construction of th.4.14 shows that X can be represented as a function module with base space K_n and one-dimensional components. (the details are left to the reader; note that $|f|[s] = |f(s)|$, where $|f|$ is the norm resolution of f).

One might suspect that for maximal function module representations the component spaces must have trivial centralizer. This is not necessarily the case as the following example shows:

<u>Counterexample</u>: There exists a real function module X in $\prod^{\infty}_{k \in [0,1]} X_k$ such that (i) $Z(X) = \{M_h \mid h \in C[0,1]\}$ (so that the identical representation of X is maximal)

(ii) $X_k = l_2^1$ ($= \mathbb{R}^2$, provided with the norm $\|(a,b)\| :=$

$|a|+|b|$) for every $k \in [0,1]$ so that all X_k have two-dimensional centralizer

Proof: Let X be the real Banach space

$\{(f,g) \mid f,g\colon [0,1] \to \mathbb{R},\ f \in C_{\mathbb{R}}([0,1]),\ g \in c_o[0,1]\} \subset \prod\limits^{\infty}_{k\in[0,1]} X_k$

(with $X_k := l^1_2$ for every k and $g \in c_o[0,1]$ iff $\{k \mid |g(k)| \geq \varepsilon\}$ is finite for every $\varepsilon > 0$).

It is an easy exercise to prove that X is a function module in $\prod\limits^{\infty}_{k\in[0,1]} X_k$ and it remains to show that $Z(X) \subset \{M_h \mid h \in C[0,1]\}$ ("$\supset$" is always valid).

For $T \in Z(X)$ we have $T = \prod T_k$ for a suitable family $(T_k)_{k\in[0,1]}$, $T_k \in Z(l^1_2)$ for every k (prop. 4.7(iv)). The operators in $Z(l^1_2)$ are represented by the matrices $\begin{pmatrix} a+b & a-b \\ a-b & a+b \end{pmatrix}$ for $a,b \in \mathbb{R}$ (this follows from $l^1_2 \cong l^\infty_2 \cong C\{0,1\}$) so that, for suitable functions $\alpha, \beta\colon [0,1] \to \mathbb{R}$,

$$T(f,g) = ((\alpha+\beta)f + (\alpha-\beta)g, (\alpha-\beta)f + (\alpha+\beta)g) \quad (\text{all } (f,g) \in X).$$

Since $(0, \chi_{\{k\}}) \in X$ it follows that $(\alpha-\beta)\chi_{\{k\}} \in C[0,1]$ for every k so that $\alpha-\beta = 0$ and consequently $T = M_{\alpha+\beta}$. $(\underline{1},0) \in X$ implies that $\alpha+\beta \in C[0,1]$. □

Note: It follows from the results in [13] that $X \hat{\otimes}_\varepsilon \dots \hat{\otimes}_\varepsilon X$ (the tensor product of n copies of X) has a maximal function module representation with base space $[0,1]^n$ and component spaces $l^1_2 \hat{\otimes}_\varepsilon \dots \hat{\otimes}_\varepsilon l^1_2 \cong l^\infty_{2^n}$ so that there are even maximal function module representations for which the centralizers of the component spaces are 2^n-dimensional (cf. also the note on p. 133).

There are two possible ways of regarding the operators in Z(X) as multiplications by continuous functions. One way is to choose a maximal function module representation of X, the other is the assertion of the Dauns-Hofmann type theorem 3.13(ii): the operators in Z(X) correspond to the structurally continuous functions on E_X. We will see at once how these two representations are related (prop. 4.20).

Let X be a Banach space and R a maximal function module representation of X. Without loss of generality we may assume that X is a function module in $\prod^{\infty}_{k\in K_X} X_k$. We define $K_X^* := \{k \mid k \in K_X,\ X_k \neq \{0\}\}$.

(Warning: Since there are in general several ways to represent X as a function module with base space K_X, the set K_X^* does not only depend on K_X but also on the representation of X as a function module. However, two such maximal representations on K_X are equivalent so that, if K_X^* and $(K_X^*)^{\sim}$ have been defined using different representations, there is a homeomorphism $t: K_X \to K_X$ such that $t(K_X^*) = (K_X^*)^{\sim}$. In particular, K_X^* is uniquely determined up to homeomorphism.)

4.19 Proposition: With K_X^* and K_X as above we have $K_X = \beta K_X^*$

Proof:

Let $\alpha: K_X^* \to \mathbb{K}$ be a bounded continuous function. We extend α by $\alpha(k) := 0$ for $k \in K_X \setminus K_X^*$ and claim that $M_\alpha X \subset X$.

For $x \in X$ and $\varepsilon > 0$ the set $K_\varepsilon := \{k \mid k \in K_X, \|x(k)\| \geq \varepsilon\}$ is closed and contained in K_X^*. Let h_ε be a continuous extension of $\alpha|_{K_\varepsilon}$ to all of K_X such that $\|h_\varepsilon\| \leq \|\alpha\|$. Since $\|\alpha x - h_\varepsilon x\| \leq 2\varepsilon\ \|\alpha\|$ and $h_\varepsilon x \in X$ it follows that $\alpha x \in X^- = X$. Thus $M_\alpha \in Z(X)$ (prop. 4.11(iii)) so that there is a function $h \in CK_X$ such that $M_h = M_\alpha$. Clearly $\alpha|_{K_X^*} = h|_{K_X^*}$ so that h is a continuous extension of α to all of K_X.

K_X^* is dense in K_X by 4.1(iv). This proves that $K_X = \beta K_X^*$. □

Note: This proposition shows that K_X must be a "large" space if there are points in K_X for which the component spaces are trivial. Another consequence is the following necessary condition concerning the existence of function modules: If K is a compact Hausdorff space and K^* a dense subset, then there exists a function module X with base space K such that $K_X = K$ and $K_X^* = K^*$ only if $\beta K^* = K$.

For example, for every Banach space X such that $K_X \cong [0,1]$ we necessarily have $K_X^* = K_X$.

<u>4.20 Proposition</u>: Let $\pi : E_X \to K_X$ be the mapping $p \mapsto k$ for $p \in E_{X_k}$ (see th. 4.5). Then a bounded function $a : E_X \to \mathbb{K}$ is structurally continuous iff there is an $h \in CK_X$ such that $a = h \circ \pi$

<u>Proof</u>: Suppose that a is structurally continuous. Since $Z(X) = \{M_h \mid h \in CK_X\}$ it follows from th. 3.13(ii) that there is an $h \in CK_X$ such that $a_{M_h} = a$. It is clear from th. 4.5 that $a_{M_h} = h \circ \pi$.
Conversely, for $h \in CK_X$, we have $M_h \in Z(X)$ so that a_{M_h} $(= h \circ \pi)$ must be structurally continuous. □

<u>C. Applications of the characterization theorems</u>

The fact that the operators in Z(X) behave as multiplication operators has a number of important consequences concerning the structure of Z(X).

<u>4.21 Proposition</u>: Z(X) is closed with respect to the topology $\mathfrak{T}_{ex}$ on B(X) (cf. prop. 4.11(iv)). In particular, Z(X) is closed with respect to the weak operator topology

<u>Proof</u>: This follows at once from prop. 4.11 and the fact that X has a maximal function module representation. □

<u>Note</u>: It is clear that Mult(X) is $\mathfrak{T}_{ex}$-closed so that, for real scalars, the assertion is obvious (recall that Z(X) = Mult(X) in this case).

It is well-known that CK is neither smooth nor strictly convex if K contains more than one point and that CK is reflexive only if K is finite. The following propostition asserts that function modules have similar properties.

4.22 Proposition: Let $(L,(Y_l)_{l\in L},Y)$ be a function module.

(i) If L contains more than one point, then Y is neither smooth nor strictly convex; also, Y has a nontrivial M-ideal

(ii) Suppose that L contains a family $(O_i)_{i\in I}$, where the $(O_i)_{i\in I}$ are nonvoid disjoint open subsets. Then Y contains a subspace which is isometrically isomorphic to $c_o(I) :=$

$\{\alpha \mid \alpha: I \to \mathbb{K}, \{i \mid |\alpha(i)| \geq \varepsilon\}$ is finite for every $\varepsilon > 0\}$.

It follows that Y contains a copy of c_o if L is infinite (so that Y cannot be reflexive in this case)

Proof:

(i) Suppose that L contains more than two points. It follows easily from 4.1(i)(iii)(iv) that there are $y_1, y_2 \in Y$ such that $\|y_1\| = \|y_2\| = 1$ and $y_1(l) = 0$ or $y_2(l) = 0$ for every $l\in L$. Thus $\|ay_1 + by_2\| = \max\{|a|, |b|\}$ for $a,b \in \mathbb{K}$ so that Y cannot be strictly convex.

If l_i is a point of L such that $\|y_i(l_i)\| = 1$ and $p_{l_i} \in (Y_{l_i})'$ a support functional for $y_i(l_i)$ $(i=1,2)$, then $p_\lambda : y \mapsto \lambda p_{l_1}(y(l_1)) + (1-\lambda)p_{l_2}(y(l_2))$ supports $\frac{1}{2}(y_1 + y_2)$ for every $\lambda \in [0,1]$, i.e. Y is not smooth. □

Further $\{y \mid y \in Y, y(l_1) = 0\}$ is a nontrivial M-ideal of Y (cor. 4.10(i)).

(ii) For $i \in I$ we choose $y_i \in Y$ such that $\|y_i\| = 1$ and $y_i|_{K \setminus O_i} = 0$ (this is possible because of 4.1(i)(iii)(iv)).

Clearly $\overline{\text{lin}}\{y_i \mid i \in I\}$ is isometrically isomorphic to $c_o(I)$.

4.23 Corollary: Suppose that X is smooth or strictly convex or that X contains no nontrivial M-ideals; then Z(X) is trivial

Proof: Z(X) is trivial iff K_X contains only one point. □

4.24 Corollary: Z(X) is finite dimensional for every Banach space which contains no copy of c_o (in particular this is true for every reflexive space)

Proof: K_X is infinite iff $Z(X)$ is infinite-dimensional. □

4.25 Corollary: $C(X)$ is finite-dimensional for every reflexive space X

Proof: X' is also reflexive so that $Z(X')$ is finite-dimensional. Thus $\mathbb{P}_M(X')$ and consequently $\mathbb{P}_L(X)$ are finite (see prop. 3.15 and prop. 1.5) so that $C(X)$ is finite-dimensional. □

We have already proved some results concerning the relations between $Z(X)$ and the M-ideals of X (prop. 3.14, prop. 3.16). The following proposition continues these investigations.

4.26 Proposition: Let X be a Banach space and J an M-ideal of X. Then $(T(J))^-$ is also an M-ideal for every $T \in Z(X)$

Proof: $(T(J))^-$ is a closed subspace and, since $Z(X)$ is commutative, a $Z(X)$-module. By prop. 4.9(ii) it remains to show that $\tilde{J}_k :=$ $\{x(k) \mid x \in (T(J))^-\}$ is an M-ideal in X_k for every $k \in K_X$ (we regard X as a function module in $\prod\limits^{\infty}_{k \in K_X} X_k$ as in th. 4.14). We write $T = M_h$ for a suitable function $h \in CK_X$ and claim that $\tilde{J}_k = \{0\}$ if $h(k) = 0$ and $\tilde{J}_k = J_k$ $(:= \{x(k) \mid x \in J\})$ if $h(k) \neq 0$. The case $h(k) = 0$ is trivial. For $h(k) \neq 0$ choose $g \in CK$ such that $(gh)(k)=1$. Thus for $x \in J$ we have $gx \in J$ so that $x(k) = (hgx)(k) = T(gx)(k) \in \tilde{J}_k$. This proves that $J_k \subset \tilde{J}_k$ so that $\tilde{J}_k = J_k$ ("$\supset$" is easily verified).□

Note: A similar result is valid for operators T in $C(X)$ and L-summands J. In this case $T^{-1}(J)$ is also an M-summand ([9], Satz D4). This has no analogue for operators in $Z(X)$ and M-ideals; even the kernel of such an operator need not be an M-ideal:

> Consider the counter-example on p.101 and the operator M_h for $h(s) := s$. $M_h(f,g) = 0$ iff $f = 0$ and $g = b\,\chi_{\{0\}}$ for a suitable $b \in \mathbb{R}$ so that $\ker M_h = \mathrm{lin}\,(0, \chi_{\{0\}})$.

$\{(f,g)(0) \mid (f,g) \in \ker M_h\} = \{(0,b) \mid b \in \mathbb{R}\}$ is not an M-ideal in l_2^1 so that $\ker M_h$ cannot be an M-ideal (prop. 4.9(ii)).

However, similarly to the preceding proof it can be shown that $T^{-1}(J)$ is an M-ideal provided that the sets $\{k \mid x(k) \in J_k\}$ are closed for every $x \in X$. In particular, $\ker T$ is an M-ideal if the functions $k \mapsto \|x(k)\|$ are continuous on K_X.

5. M-Structure of some classes of Banach spaces

We already noted that it is important for our investigations in part II to have sufficient information on the M-sturcture properties of a given Banach space X (of particular interest is a maximal function module representation). The aim of this chapter is to provide us with a number of examples where such information can be deduced from geometric, topological, or algebraic properties of the space.

We will see in part II that in view of the generalizations of the Banach-Stone theorem those Banach spaces behave "well" for which the centralizer is small. Sections A and B are devoted to some examples where the centralizers are one-dimensional or, more generally, finite dimensional.

In section C it is shown that for dual spaces X' the centralizer is completely determined by the L-summands of X. This has a number of consequences for maximal function module representations of X'.

Finally, in section D, we collect together some results concerning the M-structure of B^*-algebras.

A. Banach spaces for which the centralizer is one-dimensional

In the following proposition we restate some results which have been obtained in the previous chapters.

5.1 Proposition: Let X be a nonzero Banach space. Each of the following conditions implies that Z(X) is one-dimensional:

(i) X is smooth

(ii) X is strictly convex

(iii) X has no nontrivial M-ideal

(iv) X is reflexive and all M-summands of X are trivial

(v) $Z(X_{\mathbb{R}})$ is one-dimensional

(vi) $X \not\cong l_2^\infty$, and there exists a nontrivial L-projection on X

(vii) $X \not\cong l_2^\infty$, and X' contains a nontrivial M-ideal

Proof: (i),(ii), and (iii) have been proved in cor. 4.23. (iv) follows from (iii) and the fact that every M-ideal in a reflexive space is an M-summand by prop. 1.5 (or by prop. 2.2(ii)), and (v) is a consequence of th. 3.13(e). If X satisfies (vi) or (vii), then all M-ideals in X must be trivial by cor. 1.14 so that, by (iii), Z(X) is one-dimensional. □

Notes: 1. It can be shown that smooth spaces and strictly convex spaces have no nontrivial M-ideals so that (i) and (ii) are also a consequence of (iii). The case of strictly convex spaces is considered in cor. 6.8.
If X is smooth and J is an M-ideal of X then it is easily seen that all functionals on X which attain their norms are contained either in J^π or in $(J^\pi)^\perp$ so that, by the Bishop-Phelps theorem ([36], p.3) J^π or $(J^\pi)^\perp$ must be all of X'. Thus J = {0} or J = X.

2. By (iv) a reflexive space has a trivial centralizer if $\mathbb{P}_M(X)$ is trivial. The converse is also true by prop.3.15 since $Z(X) \cong C(K_X)$ is finite-dimensional for reflexive spaces (cor. 4.24) so that there are nontrivial idempotents if Z(X) is nontrivial.

The M-structure properties of such spaces are as follows:-

5.2 Proposition: Let X be a Banach space such that Z(X) is one-dimensional. Then all M-summands of X are trivial and the trivial representation of X (in $\prod_{k\in K}^\infty X_k$ with K = {1} and X_1 = X) is a maximal function module representation

Proof: see prop. 3.15 and the example on p.91 . □

Note: Nothing can be said in general about $\mathbb{P}_L(X)$ and the M-ideals of X (cf. the note on p. 72).

B. Banach spaces for which the centralizer is finite-dimensional

We will say that a Banach space is M-finite if Z(X) is finite-dimensional. It is clear from prop. 4.7(iv) that $Z(\prod_{i=1}^{n}{}^{\infty} X_i) \cong \prod_{i=1}^{n}{}^{\infty} Z(X_i)$ so that finite products of M-finite Banach spaces are also M-finite. In particular, $\prod_{i=1}^{n}{}^{\infty} X_i$ is M-finite if the centralizers of the X_i are one-dimensional (we will see at once that, conversely, every M-finite Banach space has this form). Examples of M-finite Banach spaces can thus be built up by the examples considered in section A.
By cor. 4.24 every reflexive space is M-finite (more generally: every space which does not contain c_o). It can be shown that every M-summand in an M-finite space and the ε-tensor product of two such spaces are also M-finite (the proof of the first assertion is elementary; the second follows from the fact that $Z(X \hat{\otimes}_\varepsilon Y) \cong Z(X) \hat{\otimes}_\varepsilon Z(Y)$ for a class of Banach spaces which contains the M-finite spaces: [13]).

5.3 Proposition: Let X be a nonzero Banach space. Then the following are equivalent:

a) X is M-finite

b) X is isometrically isomorphic to a finite product $\prod_{i=1}^{n}{}^{\infty} X_i$, where $Z(X_i) = \mathbb{K}\,\mathrm{Id}$ for $i=1,\ldots,n$

c) X contains only a finite number of M-summands, and every minimal nonzero M-summand has one-dimensional centralizer

Proof:
"a $\Rightarrow$ c": $Z(X) \cong C(K_X)$ has only a finite number of idempotents so that $\mathbb{P}_M(X)$ must be finite (prop. 3.15). Let J be an M-summand of X with

associated M-projection E. It is straightforward to show that $T\circ E \in \mathrm{Mult}(X)$ for $T \in \mathrm{Mult}(J)$ (see th. 3.3) so that $Z(J)\circ E \subset Z(X)$ (in the complex case note that $Z(J)\circ E = (Z(J_{\mathbb{R}}) + iZ(J_{\mathbb{R}}))\circ E \subset Z(X_{\mathbb{R}}) + iZ(X_{\mathbb{R}}) = Z(X)$). Thus $T \mapsto T\circ E$ maps $Z(J)$ isometrically onto a subalgebra of $Z(X)$. If J is minimal and nonzero, then this subalgebra cannot contain nontrivial idempotents so that it must be one-dimensional.

"c ⇒b": Let $X_1,\ldots,X_n$ be the collection of the nonzero minimal M-summands of X. We claim that $\prod_{i=1}^{n\,\infty} X_i \cong X$.

The mapping $\omega : \prod_{i=1}^{n\,\infty} X_i \to X$, $(x_1,\ldots,x_n) \mapsto x_1+\ldots+x_n$, is linear and onto since $X_1+\ldots+X_n = X$ (if not, $(X_1+\ldots+X_n)^{\perp}$ would contain a minimal nonzero M-summand different from $X_1,\ldots,X_n$, a contradiction). Since the $X_1,\ldots,X_n$ have pairwise intersection $\{0\}$ (as an easy consequence of the minimality) we have $E_iE_j = 0$ (for $i\neq j$) for the corresponding M-projections so that, for $x_i \in X_i$ $(i=1,\ldots,n)$,

$$\begin{aligned}\|x_1+\ldots+x_n\| &= \max\{\|E_1(x_1+\ldots+x_n)\|,\|(Id-E_1)(x_1+\ldots+x_n)\|\}\\ &= \max\{\|x_1\|,\|x_2+\ldots+x_n\|\} = \ldots = \max\{\|x_1\|,\ldots,\|x_n\|\}.\end{aligned}$$

Thus ω is an isometry.

"b ⇒a": This follows at once from $Z(\prod_{i=1}^{n\,\infty} X_i) \cong \prod_{i=1}^{n\,\infty} Z(X_i)$ (see prop.4.7 (iv)). □

5.4 Proposition: Let X be an M-finite Banach space (by prop. 5.3 we may assume that $X = \prod_{i=1}^{n\,\infty} X_i$, where $\dim Z(X_i) = 1$ for $i=1,\ldots,n$; we will regard the X_i as subspaces of X).

(i) M-ideals: The M-ideals of X are the spaces $J_1+\ldots+J_n$, where J_i is an M-ideal in X_i for every $i\in\{1,\ldots,n\}$.

(ii) M-summands: The M-summands of X are precisely the sums $J_1+\ldots+J_n$ where $J_i = \{0\}$ or $J_i = X_i$ for $i=1,\ldots,n$.

(iii) L-summands: If $X \ncong l_2^{\infty}$ and $n \geq 2$, then all L-summands of X are trivial.

(iv) Centralizer: The operators in $Z(X)$ are precisely the operators $(x_1,\ldots,x_n) \mapsto (a_1x_1,\ldots,a_nx_n)$ where $a_1,\ldots,a_n \in \mathbb{K}$.

(v) Maximal function module representation: The identical representation of X in $\prod_{i=1}^{n}{}^{\infty} X_i$ is a maximal function module representation with base space $\{1,\ldots,n\}$ and componenet spaces $X_1,\ldots,X_n$.

Proof: This follows easily from the results of chapter 4 (prop. 4.7, prop. 4.9). □

If we collect together spaces which are isometrically isomorphic we may write every M-finite Banach space X as $\prod_{\rho=1}^{r}{}^{\infty} \tilde{X}_\rho^{n_\rho}$ (Y^n denotes the product of n copies of Y, provided with the supremum norm) where the $\tilde{X}_1,\ldots,\tilde{X}_r$ are pairwise not isometrically isomorphic and $Z(\tilde{X}_\rho) = \mathbb{K}\,\mathrm{Id}$ for $\rho=1,\ldots,r$. Such a decomposition will be called a canonical M-decomposition of X, and we will say that the $n_1,\ldots,n_r$ are the M-exponents of X. The following lemma shows that canonical M-decompositions and M-exponents are uniquely determined up to rearrangement and isometric isomorphism.

5.5 Lemma: Suppose that $X \cong \prod_{\rho=1}^{r}{}^{\infty} \tilde{X}_\rho^{n_\rho}$ and $X \cong \prod_{\rho=1}^{\tilde{r}} \tilde{Y}_\rho^{m_\rho}$ are canonical M-decompositions of the M-finite Banach space X. Then there exists a bijection $t:\{1,\ldots,r\} \to \{1,\ldots,\tilde{r}\}$ such that $n_\rho = m_{t(\rho)}$ for $\rho=1,\ldots,r$ and a family of isometric isomorphisms $S_\rho:\tilde{X}_\rho \to \tilde{Y}_{t(\rho)}$

Proof:

We will use properties of maximal function module representations (cor 4.17). The isometric isomorphisms $X \cong \prod_{\rho=1}^{r}{}^{\infty} \tilde{X}_\rho^{n_\rho}$ and $X \cong \prod_{\rho=1}^{\tilde{r}}{}^{\infty} \tilde{Y}_\rho^{m_\rho}$ can be regarded as two maximal function module representations of X (with base spaces $\{1,\ldots,n_1\} \dot{\cup} \ldots \dot{\cup} \{1,\ldots,n_r\}$ and $\{1,\ldots,m_1\} \dot{\cup} \ldots \dot{\cup} \{1,\ldots,m_{\tilde{r}}\}$ and with component spaces $\underbrace{\tilde{X}_1,\ldots,\tilde{X}_1}_{n_1}, \ldots, \underbrace{\tilde{X}_r,\ldots,\tilde{X}_r}_{n_r}$ and $\underbrace{\tilde{Y}_1,\ldots,\tilde{Y}_1}_{m_1}, \ldots, \underbrace{\tilde{Y}_r,\ldots,\tilde{Y}_r}_{m_{\tilde{r}}}$). These function module representations are equivalent by th. 4.16(ii) which proves in particular that both representations contain the same number of essentially different component spaces (i.e. $r = \tilde{r}$) and that the $\tilde{X}_1,\ldots,\tilde{X}_r$ are (up to rearrangement and isometric isomorphism) just the spaces $\tilde{Y}_1,\ldots,\tilde{Y}_{\tilde{r}}$. □

Examples:

1. Suppose that $Z(X) = \mathbb{K}\,\mathrm{Id}$. In this case $X = X^1$ is a canonical M-decomposition of X so that $r = 1$, $X_1 = X$, $n_1 = 1$.
2. $C(\{1,\ldots,n\})$ is M-finite with canonical M-decomposition $\mathbb{K}^n$ (i.e. $r=1$, $X_1 = \mathbb{K}$, $n_1 = n$).

Remark: If we regard the spaces X such that $Z(X) = \mathbb{K}\,\mathrm{Id}$ as "prime" elements then the preceding result can be restated by saying that M-finite spaces admit unique factorizations into powers of prime elements. We mention some further properties of the class of M-finite spaces which are also similar to properties of the integers (X, Y, Z M-finite spaces):

- we define $X|Y$ iff X is isometrically isomorphic to an M-summand of Y; then we have $X|Y$ and $Y|X$ iff $X \cong Y$, and $X|Y$ and $Y|Z$ imples $X|Z$
- X and Y admit a "greatest common devisor" $X\wedge Y$ and a "lowest common multiple" $X\vee Y$; we have $(X\wedge Y)\times(X\vee Y) \cong X\times Y$ (the products provided with the supremum norm)
- suppose that X and Y have canonical M-decompositions $\prod_{\rho=1}^{r}{}^{\infty}\tilde{X}_\rho^{n_\rho}$ and $\prod_{\rho=1}^{\tilde{r}}{}^{\infty}\tilde{Y}_\rho^{m_\rho}$; then $X|Y$ iff there exists a mapping $\omega : \{1,\ldots,r\} \to \{1,\ldots,\tilde{r}\}$ (which is necessarily injective) such that $\tilde{X}_\rho \cong \tilde{Y}_{\omega(\rho)}$ and $m_{\omega(\rho)} \leq n_\rho$ (all $\rho \in \{1,\ldots,r\}$)
- suppose that X is prime (i.e. $Z(X) = \mathbb{K}\,\mathrm{Id}$); then $X|Y\times Z$ iff $X|Y$ or $X|Z$.

The proofs are omitted. They can easily be obtained from the preceding results.

C. Dual Banach spaces

Let X be a nonzero Banach space. In this section we will investigate the M-structure properties of X'.

5.6 Theorem: Every M-summand H of X' is weak*-closed

Proof:

It suffices to show that $H \cap \{ p \mid p \in X', \|p\| \le 1\}$ is weak*-closed. Suppose that q is a point in the weak*-closure of this set. We decompose q as $q = p_1 + p_2$, where $p_1 \in H$ and $p_2 \in H^\perp$, and we will prove that $p_2 = 0$ (so that, since $\{ p \mid p \in X', \|p\| \le 1\}$ is weak*-closed, $q = p_1 \in H \cap \{p \mid p \in X', \|p\| \le 1\}$).

Suppose that $\|p_2\| > 0$. We choose $a > 0$ such that $a\|p_2\| \ge \|p_1\| + 1$ and $\varepsilon > 0$ such that $(a+2)\varepsilon < \|p_2\|$. For suitable $x \in X$, $p \in H$ such that $\|x\|, \|p\| \le 1$ we have $|p_2(x)| \ge \|p_2\| - \varepsilon$ and $|p(x) - q(x)| \le \varepsilon$. It follows that

$$
\begin{aligned}
a\|p_2\| &= \max\{\|p - p_1\|, a\|p_2\|\} \quad (\text{since } \|p - p_1\| \le \|p_1\| + 1) \\
&= \|p - p_1 + ap_2\| \\
&\ge |(p - p_1 + ap_2)(x)| \\
&= |(p - p_1 - p_2)(x) + (a+1)p_2(x)| \\
&\ge (a+1)|p_2(x)| - |(p - p_1 - p_2)(x)| \\
&\ge (a+1)(\|p_2\| - \varepsilon) - \varepsilon
\end{aligned}
$$

in contradiction to the choice of a and ε. □

As a corollary we obtain

5.7 Theorem:

(i) Every M-summand in X' is the annihilator of an L-summand in X

(ii) $E \mapsto E'$ is an isomorphism of Boolean algebras from $\mathbb{P}_L(X)$ onto $\mathbb{P}_M(X')$; in particular, $\mathbb{P}_M(X')$ is a complete Boolean algebra

(iii) $T \mapsto T'$ is an isometric isomorphism of B^*-algebras from $C(X)$ onto $C_\infty(X')$

(recall that $C(X)$ and $C_\infty(X')$ are CK-spaces and thus B^*-algebras in a natural way)

Proof:

(i) Let H be an M-summand of X'. We define

$$J := \{x \mid x \in X,\ p(x) = 0 \text{ for every } p \in H\}$$
$$J^{\perp} := \{x \mid x \in X,\ p(x) = 0 \text{ for every } p \in H^{\perp}\}$$

(so that $J^{\pi} = H$ by th. 5.6). Similarly to the proof of prop.2.2 $a \Leftrightarrow b$ it can be shown that J and $J^{\perp}$ are complementary L-summands.

(ii) For $F \in \mathbb{P}_M(X')$ we define J as in (i) for H:=range F. Let E be the L-projection with range J. Then (Id - E)' and F are M-projections with range H so that, by lemma 1.4(ii), F = (Id-E)'. Hence $E \mapsto E'$ is onto. The other assertions are obvious.

(iii) It follows immediately from (ii) that $T \mapsto T'$ is an isometric algebra homomorphism from C(X) onto $C_{\infty}(X')$. It is further clear that, for $\sum_{i=1}^{n} a_i E_i \in \operatorname{lin}\mathbb{P}_L(X)$,

$$\left(\sum_{i=1}^{n} \overline{a}_i E_i\right)' = \sum_{i=1}^{n} \overline{a}_i E_i' = \left(\sum_{i=1}^{n} a_i E_i'\right)^*$$

so that, by continuity, $(T^*)' = (T')^*$ for every $T \in C(X)$. □

Remarks: 1. It is well-known that a space can have preduals which are not isometrically isomorphic. The preceding theorem asserts that the structure of the L-summands must be the same for all such preduals: $X' \cong Y' \Rightarrow \mathbb{P}_L(X) \cong \mathbb{P}_L(Y)$.

2. The definition "a closed subspace J of X is an L-ideal if J^{π} is an M-summand" which is analogous to the definition of M-ideals is of no use since by th. 5.7(i) every "L-ideal" is an L-summand.

5.8 Lemma: Let X be a real Banach space and $T \in Z(X')$ such that $0 \le T \le Id$ (recall that $T \ge 0$ iff $a_T \ge 0$). Then there is an $F \in \mathbb{P}_M(X')$ such that $(Fp)(x) = \lim_n (T^n p)(x)$ for every $p \in X'$, $x \in X$

Proof: $0 \le T \le Id$ implies that $0 \le T' \le Id$ in $C(X'')$ (this follows from th. 3.12(c) and the fact that the positive elements in Z(X) are just the squares). Thus $(T')^n$ converges pointwise to an L-projection E on X'' (prop. 3.11(ii)). In particular $\lim (T^n p)(x)$ exists for

$p \in X'$ and $x \in X$. Define $F: X' \to X'$ by $(Fp)(x) := \lim (T^n p)(x)$.
It is clear that F is well-defined, linear and continuous and that, since the T^n have this property, satisfies the condition for M-boundedness with $\lambda = 1$. Accordingly $F \in Z(X')$.

$\nu(F)$ and the $\nu(T^n)$ are continuous functions on $K_{X'}$ (where ν is an isometric B^*-algebra isomorphism from $Z(X')$ onto $C(K_{X'})$) and we will prove that range $\nu(F) \subset \{0,1\}$ so that F must be a projection in $Z(X')$, i.e. an M-projection (prop. 3.15).

It suffices to prove that $\nu(F)(A) \subset \{0,1\}$ for a dense subset A of $K_{X'}$, and we will show that $A := A_1 \cup A_2$

$$\begin{pmatrix} A_1 := \{k \mid k \in K_{X'},\ \nu(T)|_U = 1 \text{ for a neighbourhood } U \text{ of } k\} \\ A_2 := \{k \mid k \in K_{X'},\ 0 \le \nu(T)(k) < 1\} \end{pmatrix}$$

has this property; note that $0 \le \nu(T) \le 1$ so that A is dense in $K_{X'}$.

For $k \in A_1$ let U be an open neighbourhood of k such that $\nu(T)|_U = 1$. For $p \in X'$ such that $p|_{K_{X'} \setminus U} = 0$ (we regard X' as a function module with base space $K_{X'}$ as in th. 4.14; note that $Tp = \nu(T)p$ in this representation) we have $Tp = p$ so that $Fp = p$. Thus $\nu(F)(1) = 1$ at every point of U for which the component space is nonzero. These points are dense in U so that $\nu(F)(k) = 1$.

For $k \in A_2$ choose an open neighbourhood U of k such that $\|\nu(T)|_U\| < \delta$ for a suitable $\delta < 1$. But then, for $p \in X'$ such that $p|_{K_{X'} \setminus U} = 0$ it follows that $\|T^n p\| \le \delta^n \|p\|$ so that $Fp = 0$. Thus $\nu(F)$ is zero on a dense subset of U (at every point for which the component space is nonzero), i.e. $\nu(F)|_U = 0$. □

5.9 Theorem: $Z(X') = C_\infty(X')$

Proof:

Suppose first that $\mathbb{K} = \mathbb{R}$. For $T \in Z(X')$ and $\varepsilon > 0$ we choose $S_1, .., S_k \in Z(X')$ and numbers $b_1, ..., b_k \in \mathbb{R}$ such that $\|T - \sum_{i=1}^{k} b_i S_i^n\| \le \varepsilon$ for every $n \in \mathbb{N}$ and $0 \le S_i \le Id$ (cf. the similar construction on p. 68). The $(S_i^n p)(x)$ converge to $(F_i p)(x)$ where the F_i are M-projections (lemma 5.8) for $p \in X'$, $x \in X$ so that $\| T - \sum_{i=1}^{k} b_i F_i \| \le \varepsilon$. This proves that $T \in \overline{\mathrm{lin}\,\mathbb{P}_M(X')} = C_\infty(X')$.

In the complex case we have $Z(X'_{\mathbb{R}}) = C_\infty(X'_{\mathbb{R}})$ by the first part of the proof (note that $(X')_{\mathbb{R}} = (X_{\mathbb{R}})'$) so that $Z(X') = Z(X'_{\mathbb{R}}) + iZ(X'_{\mathbb{R}}) \subset C_\infty(X')$.

The reverse inclusion is obvious since $\mathbb{P}_M(X') \subset Z(X')$. □

<u>5.10 Corollary</u>: $K_{X'}$ is an extremally disconnected space

<u>Proof</u>: By th. 5.7 and th. 5.9 we have $C(K_{X'}) \cong Z(X') \cong C(X) \cong C(\Omega_X)$ and Ω_X is extremally disconnected. The assertion is thus a consequence of the Banach-Stone theorem. □

Let $\nu : Z(X') \to C(\Omega_X)$ be defined by $\omega_X^{-1} \circ P$, where P is the inverse of the isomorphism $T \mapsto T'$ (from $C(X)$ onto $Z(X')$; note that ν is an isometric isomorphism of B^*-algebras and that $(\omega_X \circ \nu(T))' = T$ for every $T \in Z(X')$. By th. 4.14 there exists a function module representation $\rho : X' \to \prod_{k \in \Omega_X}^{\infty} Y_k$ of X' such that $\rho(Tp) = \nu(T)\,\rho(p)$ for every $T \in Z(X')$, $p \in X'$. For simplicity we will identify X' with $\rho(X')$; note that $\omega_X^{-1}(E)p$ is just $p \circ E$ by the definition of ρ and ν (all $p \in X'$, $E \in \mathbb{P}_L(X)$).

<u>5.11 Lemma</u>: $\|p(k)\| = \inf\{\|p \circ E\| \mid E \in \mathbb{P}_L(X),\ (\omega_X^{-1}(E))(k) = 1\}$

$=: \|p(k)\|_*$ (all $p \in X'$, $k \in \Omega_X$)

<u>Proof</u>: We have $\|p(k)\| = \inf\{\|hp\| \mid h \in C(\Omega_X),\ h \ge 0,\ h(k) > 1\}$ as an easy consequence of the upper semicontinuity of $\|p(\cdot)\|$. We already noted that $\|p \circ E\| = \|\omega_X^{-1}(E)p\|$ so that $\|p(k)\| \le \|p \circ ((1+\varepsilon)E)\|$ for every $\varepsilon > 0$ and every $E \in \mathbb{P}_L(X)$ such that $(\omega_X^{-1}(E))(k) = 1$. This proves $\|p(k)\| \le \|p(k)\|_*$.

For continuous $h : \Omega_X \to [0,\infty[$ such that $h(k) > 1$ choose a clopen neighbourhood U of k such that $h|_U \ge \chi_U$ (recall that Ω_X is extremally disconnected). We have $\chi_U = \omega_X^{-1}(E)$ for a suitable $E \in \mathbb{P}_L(X)$ so that $\|hp\| \geq \|\chi_U p\| = \|p \circ E\| \geq \|p(k)\|_*$. This gives $\|p(k)\| \geq \|p(k)\|_*$. □

<u>5.12 Lemma</u>: There exists a $p \in X'$ such that $\|p \circ E\| = 1$ for every nonzero L-projection E of X

<u>Proof</u>: For $x \in X$ let E_x be the <u>carrier projection</u> of x: $E_x := \inf\{E \mid E \in \mathbb{P}_L(X) , Ex = x\}$. Let $(x_i)_{i \in I}$ be a maximal family of nonzero elements of X such that the $(E_{x_i})_{i \in I}$ are pairwise disjoint (such a family can be obtained by a standard application of Zorn's lemma). Since the family $(x_i)_{i \in I}$ is maximal we have $\sup\limits_i E_{x_i} = Id$. We claim that

α) $X \cong \prod\limits_{i \in I}{}^1 X_i \qquad (X_i := \text{range } E_{x_i})$

β) for $i \in I$ there exists $p_i \in X_i'$ such that $\|p_i \circ E\| = 1$ for every $E \in \mathbb{P}_L(X)$ such that $0 \neq E \leq E_{x_i}$

γ) $p : X \to \mathbb{K}$, $(y_i) \mapsto \sum\limits_{i \in I} p_i(y_i)$ has the claimed properties

α) Let ΣX_i be the subspace of all tuples in $\prod\limits_{i \in I}{}^1 X_i$ for which at most a finite number of components are nonzero. ΣX_i is dense in $\prod\limits_{i \in I}{}^1 X_i$, and $(y_i) \mapsto \sum\limits_{i \in I} y_i$ is a linear isometry from ΣX_i into X which has a unique extension to an isometric isomorphism from $\prod\limits_{i \in I}{}^1 X_i$ onto the L-summand $\overline{\text{lin}}(\bigcup\limits_{i \in I} X_i)$ of X. $\sup E_{x_i} = Id$ implies that $\overline{\text{lin}}(\bigcup\limits_{i \in I} X_i) = X$.

β) We define $\tilde{p}_i : \text{lin}\{Ex_i \mid E \in \mathbb{P}_L(X)\} \to \mathbb{K}$ by $\sum\limits_{j=1}^n a_j E_j x_i \mapsto \sum\limits_{j=1}^n a_j \|E_j x_i\|$. $\tilde{p}_i$ is well-defined (this follows from the fact that $\|(E + F)x_i\| = \|Ex_i\| + \|Fx_i\|$ for disjoint $E, F \in \mathbb{P}_L(X)$) and linear. For $\sum\limits_{j=1}^n a_j E_j x_i \in \text{lin}\{Ex_i \mid E \in \mathbb{P}_L(X)\} \setminus \{0\}$ we may assume that $a_j \neq 0$, $E_j \neq 0$, $E_{j_1} E_{j_2} = 0$ (all $j, j_1, j_2 \in \{1, \dots, n\}$, $j_1 \neq j_2$) so that

$$\begin{aligned} |\tilde{p}_i(\sum_{j=1}^n a_j E_j x_i)| &= |\sum_{j=1}^n a_j \|E_j x_i\| | \\ &\leq \sum_{j=1}^n |a_j| \|E_j x_i\| \\ &= \|\sum_{j=1}^n a_j E_j x_i\| \end{aligned}$$

so that $\tilde{p}_i$ is continuous with $\|\tilde{p}_i\| \leq 1$. Let p_i be a linear extension of $\tilde{p}_i$ to all of X_i such that $\|p_i\| = \|\tilde{p}_i\|$.

For $E \in \mathbb{P}_L(X)$ such that $0 \neq E \leq E_{x_i}$ we have $Ex_i \neq 0$ by the definition of E_{x_i}. It follows that $1 \geq \|p_i\| \|E\| \geq \|p_i \circ E\| \geq |p_i \circ E(Ex_i / \|Ex_i\|)| = 1$.

γ) It is obvious that p is well-defined, linear and continuous and that $\|p\| \leq 1$. For $E \in \mathbb{P}_L(X)$ such that $E \neq 0$ we have $E \circ E_{x_i} \neq 0$ for a suitable $i \in I$ (this is a consequence of $\sup_i E_{x_i} = \mathrm{Id}$). $E \circ E_{x_i} \neq 0$ implies that $Ex_i \neq 0$ so that

$$
\begin{aligned}
|(p \circ E)(Ex_i / \|Ex_i\|)| &= |p_i(Ex_i / \|Ex_i\|)| \\
&= \|Ex_i\| / \|Ex_i\| \\
&= 1.
\end{aligned}
$$

It follows that $\|p \circ E\| \geq 1$. □

<u>5.13 Theorem</u>: Let $[\rho, (K, (Y_l)_{l \in K}, \rho(X'))]$ be a maximal function module representation of X'. Then

(i) the functions $\|\rho(p)(\cdot)\|$ are continuous (and not only upper semi-continuous) for every $p \in X'$

(ii) there is a $p \in X'$ such that $\|\rho(p)(k)\| = 1$ for every $k \in K$

<u>Proof</u>: Since maximal function module representations are equivalent it suffices to prove the theorem for any maximal function module representation of X' (it is clear from def. 4.12 that, if $R_1 \approx R_2$ and R_1 satisfies (i) and (ii), then R_2 also has this property). We will consider the same representation as in the preceding lemmas (base space: Ω_X) and we will regard X' as a function module (i.e. $\rho = \mathrm{Id}$).

(i) Suppose that $k \in \Omega_X$, $p \in X'$, and that $\|p(k)\| > r$. We claim that there is a neighbourhood U of k such that $\|p(l)\| \geq r$ for $l \in U$ (i.e. $\|p(\cdot)\|$ is lower semicontinuous at k). If this is not the case there is a point k_U in each neighbourhood U of k such that $\|p(k_U)\| < r$. Choose $E_U \in \mathbb{P}_L(X)$ such that $(\omega_X^{-1}(E_U))(k_U) = 1$ and $\|p \circ E_U\| < r$ (lemma 5.11). We define $E \in \mathbb{P}_L(X)$ to be the supremum of the E_U. Since $\tilde{E} \mapsto \omega_X^{-1}(\tilde{E})$ is an order isomorphism, $\omega_X^{-1}(E)$ must be one at every k_U and thus at k so that $\|p(k)\| \leq \|p \circ E\|$ by lemma 5.11. But E is the pointwise limit of the finite suprema of the E_U (th. 1.10(i)) so that, for $\varepsilon > 0$ and $x \in X$ such that $\|x\| = 1$, there are $k_{U_1}, \dots, k_{U_n}$ such that $\|Ex - (\sup\{E_{U_i} \mid i=1,\dots,n\})(x)\| \leq \varepsilon$. Accordingly $|p(Ex)| \leq \|p\circ(\sup\{E_{U_i} \mid i=1,\dots,n\})\| + \varepsilon \leq r+\varepsilon$ so that $\|p(k)\| \leq r$ (we used the fact that $\|p\circ E_i\| \leq r$ for $i=1,\dots,n$ implies that $\|p\circ(\sup_i E_i)\| \leq r$; this is an easy exercise).

This contradiction proves that $\|p(\cdot)\|$ is lower semicontinuous and thus continuous.

(ii) This follows at once from the preceding lemmas. □

Note: Let $p \in X'$ be as in th. 5.13(ii). It follows from the fact that maximal function module representations are finer than any other representation that $\|\rho(p)(k)\| = 1$ for every function module representation of X' and every point of the base space. In particular, for every function module representation of X' all component spaces are nonzero. This is also a consequence of the simple fact that an element of a function module which vanishes at some point cannot be an extreme point of the unit ball.

However, the assertion of th. 5.13(i) is not true for every function module representation of X'. Consider, for example, the space $X := L^1[0,1]$. The multiplication operators associated with the elements of $C[0,1]$ define operators in the centralizer of $X' = L^\infty[0,1]$. Since every closed self-adjoint subalgebra of $Z(X')$ which contains Id defines a function module representation, it follows that $L^\infty[0,1]$ admits such a representation with base space $[0,1]$. We have

$$\|p(t)\| = \inf\{\|hp\| \mid h \in C[0,1],\ h \geq 0,\ h(t) > 1\} \qquad (\text{all } t \in [0,1])$$

so that it is easy to find elements $p \in L^\infty[0,1]$ such that $\|p(\cdot)\|$ is not continuous (choose, for example, a function $p \in L^\infty[0,1]$ such that $\|p(\cdot)\|$ is upper semicontinuous but not continuous; note that $\|p(t)\| = |p(t)|$ in this case for every $t \in [0,1]$).

D. B^*-algebras

In the preceding chapters we described completely the M-structure properties of C_oL:-

5.14 Proposition: Let L be a nonvoid locally compact Hausdorff space. Then
- the M-summands (M-ideals) of C_oL correspond to the clopen (closed) subsets of L
- the centralizer of C_oL contains precisely the operators M_h, $h \in C^bL$
- C_oL admits a maximal function module representation with base space βL and component spaces $\mathbb{K}_k$ (where $\mathbb{K}_k = \mathbb{K}$ for $k \in L$ and $\mathbb{K}_k = \{0\}$ for $k \in \beta L \setminus L$).

If C_oL is not isometrically isomorphic to the real space l_2^∞ (i.e. if the scalars are complex or if L contains more than two points) then all L-summands of C_oL are trivial. □

We also collected together without proof some results concerning the M-structure properties of more general B^*-algebras which we also repeat here for the sake of easy reference:-

5.15 Proposition: Let A be a B^*-algebra with unit and $X := A_{sa}$, the real Banach space of the self-adjoint elements of A. Then
- the M-ideals of A (of X) are just the (self-adjoint parts of the) closed two-sided ideals of A
- the centralizer of A (of X) contains precisely the (restrictions to X of the) multiplication operators associated with the (self-adjoint) elements of the centre of A

Proof: [3], [60], [82] □

Note: Function module representations of W^*-algebras are considered in a number of papers (see, for example, [19],[35],[38],[52],[68], [76],[86]). We note that, by th. 5.13, the norm resolutions in maximal function module representations of W^*-algebras must be continuous.

6. Remarks

This chapter contains some bibliographical notes as well as a number of supplements.

Remarks concerning chapter 1

The first systematic treatment of L-projections is due to Cunningham ([29]). Most of the results in sections A and B concerning L-summands can be traced back to this paper. Investigations of M-summands have been included for completeness, the facts contained here seem to be more or less well-known.
The proof of th. 1.12 in section C is essentially the proof of Hirsberg ([58], p. 135). The L-M-theorem and the other results of section D are special cases of more general theorems proved by the author ([8]).
Banach algebra methods are often used in the theory of Boolean algebras of projections. The proofs of our preliminary results concerning the Cunningham algebra in section C are similar to the proofs in [16], chapter 2. The Cunningham algebra of a space was first used by Alfsen and Effros ([4], chapter 4) to characterize the M-bounded operators (in the case of real scalars).
$C_\infty(X)$ has attracted less attention. We refer the reader to [34].

Supplement: L^p-summands and L^p-projections

L-projections and M-projections are special cases of a more general definition:

6.1 Definition: Let X be a Banach space, $E:X \to X$ a projection and $1 \leq p < \infty$. E is called an L^p-projection, if $\|x\|^p = \|Ex\|^p + \|x-Ex\|^p$ for every $x \in X$. A closed subspace J of X is called an L^p-summand if J is the range of a suitable L^p-projection.

Note: We will extend the definition by saying that M-projections

(M-summands) are L^∞-projections (L^∞-summands).

By the following theorem which is due to the author ([8]) it is possible to show that L^p-projections (for $1 \le p < \infty$, $p \ne 2$) behave in many respects like L-projections. A special case of (ii) has been proved by Sullivan ([85]; the case of spaces which satisfy the Clarkson inequality).

<u>6.2 Theorem</u>: Let X be a Banach space.

(i) Suppose that $p,q \in [1,\infty]$ and that $p \ne q, \{p,q\} \ne \{1,\infty\}$. Then all L^p-projections or all L^q-projections are trivial, i.e. 0 or Id (this is also true for $\{p,q\} = \{1,\infty\}$, provided that $X \not\cong l_2^\infty$; see th. 1.13)

(ii) If $p \ne 2$, then every two L^p-projections commute.

The following properties of L^p-projections and L^p-summands which are similar to the results for L-projections and L-summands in this volume are proved in [8], [9], [16], [44]:

- $\mathbb{P}_p(X)$, the collection of all L^p-projections of X, is a complete Boolean algebra ($1 \le p < \infty$, $p \ne 2$)
- if E is a projection, then $E \in \mathbb{P}_p(X)$ iff $E' \in \mathbb{P}_q(X')$ ($1 \le p \le \infty$, $1/p + 1/q = 1$)
- $E \mapsto E'$ is an isomorphism of Boolean algebras from $\mathbb{P}_p(X)$ onto $\mathbb{P}_q(X')$ ($1 \le p < \infty$, $p \ne 2$, $1/p + 1/q = 1$)
- $C_p(X) := \overline{\text{lin}}\, \mathbb{P}_p(X)$ is a commutative Banach algebra which is isometrically isomorphic to the space of continuous scalar-valued functions on the Stonean space of $\mathbb{P}_p(X)$ ($1 \le p \le \infty$, $p \ne 2$)
- if J is an L^p-summand and $T \in C_p(X)$, then $(T(J))^-$ and $T^{-1}(J)$ are also L^p-summands ($1 \le p < \infty$, $p \ne 2$)
- $T \mapsto T'$ is an isometric isomorphism from $C_p(X)$ onto $C_q(X')$ ($1 \le p < \infty$, $p \ne 2$)
- the commutator of $C_p(X)$ in $B(X)$ is identical with $C_p(X)$ iff X is an abstract L^p-space ($1 \le p < \infty$, $p \ne 2$).

Supplement: The Cunningham algebra

Up to now we have only proved those properties of the Cunningham algebra which were important in investigating M-structure properties, in particular the centralizer, of a given space.

Since $C(X) = C(X_{\mathbb{R}}) + iC(X_{\mathbb{R}})$ for complex spaces it is sufficient to consider real spaces. The following characterization theorem for the operators in $C(X)$ is due to Alfsen and Effros ([4], th. 1.11). They give an order theoretical proof; we prefer to apply the results of chapter 3 and chapter 5:

6.3 Theorem: Let X be a real Banach space and $T \in B(X)$. Then $T \in C(X)$ iff there exists an $a > 0$ such that

(*) $$\|Tx + ax\| + \|Tx - ax\| = 2a\|x\| \quad \text{for every } x \in X$$

Proof: Suppose that $S = \sum_{i=1}^{n} a_i E_i \in \operatorname{lin} \mathbb{P}_L(X)$. We may assume that the E_i are disjoint, that $\sum_{i=1}^{n} E_i = Id$, and that $a_i = 0$ if $E_i = 0$ so that $\|S\| = \max\{|a_i| \mid i=1,\dots,n\}$. A direct verification shows that (*) is satisfied for every $a \geq \|S\|$:

$$\begin{aligned} \|ax + Sx\| + \|ax - Sx\| &= \|\sum_{i=1}^{n} (a+a_i)E_i x\| + \|\sum_{i=1}^{n} (a-a_i)E_i x\| \\ &= \sum_{i=1}^{n} (a+a_i)\|E_i x\| + \sum_{i=1}^{n} (a-a_i)\|E_i x\| \\ &= 2a \sum_{i=1}^{n} \|E_i x\| \\ &= 2a\|x\| . \end{aligned}$$

Thus, for $T \in C(X) = \overline{\operatorname{lin}} \mathbb{P}_L(X)$ and $T = \lim S_n$ (all $S_n \in \operatorname{lin} \mathbb{P}_L(X)$) it follows that T satisfies (*) for every $a \geq \sup\{\|S_n\| \mid n \in \mathbb{N}\}$.

Conversely, suppose that (*) is satisfied for T. We will prove that T' satisfies the condition for M-boundedness (def. 3.2) for $\lambda = a$ so that $T' \in Z(X') = \{S' \mid S \in C(X)\}$ (th. 3.12(i), th. 5.7(iii), th. 5.9), i.e. $T \in C(X)$.

Let $p, q \in X'$, $r > 0$ be given such that $\|\pm ap - q\| \leq r$. For $x \in X$ (*) implies that
$$\begin{aligned} (p \circ T - q)(x) &= \frac{1}{2a}[(ap-q)(ax+Tx) - (ap+q)(ax-Tx)] \\ &\leq \frac{r}{2a} (\|ax+Tx\| + \|ax-Tx\|) \\ &= r\|x\| \end{aligned}$$

so that $\| p \circ T - q \| \leq r$. Thus T' is M-bounded as claimed. □

Note: It follows from the proof that we may choose $a = \|T\|$ if (*) is satisfied.

6.4 Corollary: Let Y be a closed subspace of X and $T \in C(X)$ such that $TY \subset Y$ and $T^*Y \subset Y$ (recall that $T^* := \omega_X(\overline{\omega_X^{-1}(T)})$). Then $T|_Y \in C(Y)$

Proof: If X is a real space, then the assertion follows immediately from th. 6.3. In the complex case we consider $\frac{T+T^*}{2}, \frac{T-T^*}{2i} \in C(X_{\mathbb{R}})$. We have $\frac{T+T^*}{2}\Big|_Y$, $\frac{T-T^*}{2i}\Big|_Y \in C(Y_{\mathbb{R}})$ by th. 6.3 so that $T|_Y \in C(Y)$. □

Remarks: 1. A similar result for Z(X) follows from the characterization theorem 3.13(i): If $T \in Z(X)$ and $TY \subset Y$, $T^*Y \subset Y$ (Y a closed subspace of X), then $T|_Y \in Z(Y)$.

2. Th. 6.3 also implies that limits with respect to the weak operator topology of norm bounded nets of operators in C(X) are also contained in C(X) (X a real Banach space). However, this is a special case of a general property of complete Boolean algebras of projections: the closed linear span (closure with respect to the norm topology) of such a Boolean algebra is always closed in the weak operator topology (see cor. XVII.3.17 in [40]).

Finally we note that the Stonean space Ω_X associated with $\mathbb{P}_L(X)$ is not only extremally disconnected but hyperstonean. It is not hard to prove that the functionals $p_x : C_{\mathbb{R}}(\Omega_X) \to \mathbb{R}$ defined as the unique continuous linear extensions of the mappings $\chi_O \mapsto \|(\omega_X(O))(x)\|$ (for $O \subset \Omega_X$, O clopen; cf. the construction in lemmma 5.12) are order continuous. It is clear that the family $(p_x)_{x \in X}$ separates the functions of $C_{\mathbb{R}}(\Omega_X)$.

Remarks concerning chapter 2

M-ideals were introduced by Alfsen and Effros (see [3] and chapter 2 in [4]) for the case of real spaces (which is no essential restriction since M-ideals are $\mathbb{R}$-determined). Readers who are interested in the predecessors of the theory of M-ideals are referred to chapter 6 in [3].

The characterization theorems for M-ideals (th. 2.17, th. 2.20) are proved in [3]. The original proofs are unnecessarily complicated. The more elementary proofs presented here are due to the author (we note that some ideas of these proofs, for example the assertion of th. 2.12, are already contained in an unpublished paper of R. Evans; for an alternative proof we refer the reader to [70]).

The assertions in the first part of chapter 2 are essentially the same as in chapter 2 of [4] (prop. 2.2 seems to be new).

M-ideals for special classes of spaces have been determined in [58], [70], and [82]. Of particular interest are generalizations of the fact that K(H) is an M-ideal in B(H) (H a Hilbert space): [48], [57], [61], [62], [63], [64], [79] investigate classes of Banach spaces where the subspace of compact operators is an M-ideal in the space of bounded operators and the approximation theoretical properties of this M-ideal.

Supplement: M-ideals and approximation theory

M-ideals have a number of approximation theoretical properties. The following proposition is a slight generalization of cor. 5.6 in [3]:

6.5 Proposition: Let J be a closed subspace of the Banach space X. If J satisfies the two-ball property for open balls (in particular if J is an M-ideal), then J is proximinal, i.e.

$$P_J(x) := \{ y \mid y \in J, \|x-y\| = d(x,J) \}$$

is non-empty for every $x \in X$

Proof: Consider the ball $B(x,d)$, where $d := d(x,J)$ (without loss of generality we may assume that $d > 0$). We choose $\delta < 1$ as in prop.2.18 for this ball. By definition, there is an x_1 in $J \cap B(x,d+1)$ so that, by prop. 2.18, $B(x,d+\delta) \cap B(x_1,\delta) \neq \emptyset$. Since J has the two-ball property for open balls there is an x_2 in $J \cap B(x,d+\delta) \cap B(x_1,\delta)$. Prop. 2.18 yields $B(x,d+\delta^2) \cap B(x_2,\delta^2) \neq \emptyset$ so that there exists an x_3 in $J \cap B(x,d+\delta^2) \cap B(x_2, \delta^2)$. ... . This construction provides us with a sequence $(x_n)_{n\in\mathbb{N}}$ in J such that $\|x - x_n\| \leq d+\delta^{n-1}$, $\|x_{n+1} - x_n\| \leq \delta^n$. It is clear that $\lim x_n \in P_J(x)$. □

It is now easy to prove that the $P_J(x)$ are large subsets of J if x is not contained in J:

6.6 Proposition: Let J be an M-ideal in the Banach space X (or, more generally, a closed subspace of X which satisfies the two-ball property for closed balls).

(i) For $x,y \in X$ such that $\|x-y\| < d(x,J) + d(y,J)$ we have

$P_J(x) \cap P_J(y) \neq \emptyset$

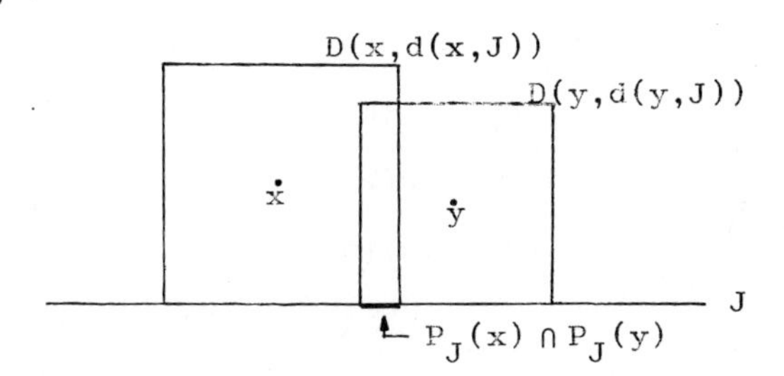

fig. 14

(ii) Suppose that $x \notin J$ and that $y \in J$, $\|y\| < d(x,J)$. Then there are

$y_1, y_2 \in P_J(x)$ such that $y = 1/2(y_1 - y_2)$

Proof:

(i) The balls $D_1 := D(x,d(x,J))$ and $D_2 := D(y,d(y,J))$ satisfy $P_J(x) = D_1 \cap J \neq \emptyset \neq D_2 \cap J = P_J(y)$ and $(D_1 \cap D_2)^o \neq \emptyset$ so that $P_J(x) \cap P_J(y) = D_1 \cap D_2 \cap J \neq \emptyset$.

(ii) We have $d(2y+x,J) = d(x,J)$ so that x and $2y+x$ satisfy the conditions of (i). Thus $P_J(x) \cap P_J(2y+x) = P_J(x) \cap (2y + P_J(x)) \neq \emptyset$ which proves (ii). □

6.7 Corollary([63]): Suppose that J is an M-ideal in X. Then

(i) $J = \operatorname{lin} P_J(x)$ for every $x \notin J$

(ii) if $P_J(x)$ is compact for any $x \notin J$, then J is finite-dimensional (and thus an M-summand by prop. 2.2(ii))

Proof:

(i) follows immediately from prop. 6.6(ii).

(ii) Suppose that $P_J(x)$ is compact, where $x \notin J$. Then $P_J(x) - P_J(x)$ is a compact subset of J with non-empty interior (this set contains $\{2y \mid y \in J, \|y\| < d(x,J)\}$) so that J must be finite-dimensional. □

6.8 Corollary: If X is strictly convex, then all M-ideals of X are trivial

Proof: Suppose that J is an M-ideal of X such that $J \neq X$. For a suitable $x \notin J$ we have $0 \in P_J(x)$ so that, since X is strictly convex, $\{0\} = P_J(x)$. It follows that $J = \text{lin}\, P_J(x) = \{0\}$. □

By a refined version of prop. 6.6 it can be shown that the map $x \mapsto P_J(x)$ satisfies the conditions of Michael's selection theorem ([73]). The proof of the following theorem depends essentially on Michael's result. For details we refer the reader to [47] and [63] .

6.9 Theorem: Let J be an M-ideal in the Banach space X. Then there exists a continuous map $f: X \to J$ such that $f(x) \in P_J(x)$ and $f(\lambda x) = \lambda f(x)$ for every $x \in X$ and every $\lambda \in \mathbb{R}$.

Note: Since there are non-complemented M-ideals (e.g. c_o in m) it is not to be expected that f can be chosen to be continuous and additive.

Remarks concerning chapter 3

Centralizers of special classes of Banach spaces and ordered linear spaces have been investigated by several authors (see the references in [4]). However, the definition of "centralizer" varies considerably. For example, in the theory of ordered linear spaces, the order centre $Z_{ord}(E)$ is the order ideal generated by the identity operator from E to E (E an ordered linear space; cf. [72],[80],[89]). For ordered linear space X which are also Banach spaces $Z_{ord}(X)$ is in general different from $Z(X)$ (see chapter 6 in [4]).

The centralizer of Banach spaces as defined in these notes was first considered by Cunningham ([30]). Alfsen and Effros ([4], chapter 4) proved that, for the case of real spaces, $T \in Z(X)$ iff $T' \in C(X')$ iff T is M-bounded iff a_T is structurally continuous. Our definition of M-boundedness for the complex case (def. 3.2) as well as th. 3.3 and the systematic development of the theory for arbitrary (i.e. real or complex) Banach spaces seem to be new.

Most of the proofs presented in this chapter are different from the proofs of Alfsen and Effros. The proof of the Dauns-Hofmann-type theorem 3.8 is essentially the same as in [43]. Generalized versions of this theorem are considered in [42] and [46].

<u>Supplement: The centralizer of tensor products</u>

Let X and Y be Banach spaces and $X\hat{\otimes}_\varepsilon Y$ their usual ε-tensor product. The problem as to how $Z(X\hat{\otimes}_\varepsilon Y)$ can be constructed from $Z(X)$ and $Z(Y)$ has been discussed in [13], [14], [88].

For $T \in Z(X)$ and $S \in Z(Y)$ the operator $T \otimes S$ is an element of $Z(X\hat{\otimes}_\varepsilon Y)$ so that $Z(X) \otimes Z(Y)$ can be regarded as a subalgebra of $Z(X\hat{\otimes}_\varepsilon Y)$. With this identification it can be proved that

- $Z(X\hat{\otimes}_\varepsilon Y)$ is the closure in the strong operator topology of $Z(X) \otimes Z(Y)$
- if X and Y have a centralizer-norming system (see def. 9.1), then $Z(X\hat{\otimes}_\varepsilon Y)$ is the norm-closure of $Z(X) \otimes Z(Y)$ (which implies that $K_{X\hat{\otimes}_\varepsilon Y} = K_X \times K_Y$ in this case).

The proofs are given for real spaces but they can easily be generalized to arbitrary spaces.

<u>Supplement: The bi-commutator of Z(X)</u>

One might suspect that as $Z(X)$ is a weakly closed B^*-algebra in $B(X)$, the bi-commutator of $Z(X)$ would be just $Z(X)$. The following counterexample (which is due to R. Evans) shows that this is not true in general:

Counterexample: There is a Banach space X such that the bi-commutator of Z(X) is strictly larger than Z(X)

Proof: Let $K := \alpha\mathbb{N}$, $X_k := \mathbb{K}$ for $k \in \mathbb{N}$ and $X_\infty := \mathbb{K}^2$, provided with the norm $\|(a,b)\| := |a| + |b|$. It is obvious that

$X := \{ (a_1, b_1, \ldots, (a,b)) \mid a_i, b_i, a, b \in \mathbb{K},\ a_i \to a, b_i \to b\}$

is a function module in $\prod^\infty_{k\in K} X_k$.

It can easily be shown that the operators of the form $\prod^\infty_{k\in K} T_k$ which leave X invariant are precisely the operators

$$(*) \qquad (a_1, b_1, \ldots, (a,b)) \mapsto (\alpha_1 a_1, \beta_1 b_1, \ldots, (\alpha a, \beta b)),$$
$$\text{where } \alpha_i \to \alpha \text{ and } \beta_i \to \beta .$$

Since an operator $(a,b) \mapsto (\alpha a, \beta b)$ belongs to $Z(X_\infty)$ iff $\alpha = \beta$ it follows from prop. 4.7 that $Z(X) = \{M_h \mid h \in CK\}$ and that the commutator of Z(X) is the space of all operators defined as in (*). Since this set is commutative and strictly larger than Z(X) it follows that Z(X) is a proper subset of its bi-commutator. □

Remarks concerning chapter 4

Function modules have a long history (cf. the reduction theory papers of von Neumann [76] and Godement [52]). They have been used at several places in functional analysis, in particular in the theory of normed algebras and approximation theory ([19] , [35] , [38] , [50] , [51] , [60] , [68] , [71] , [75] , [86]). The first investigations of the connections between function module representations and M-structure theory and the proof of the existence of a maximal functional representation (th. 4.14) are contained in [30]. Theorem 4.5 (extreme functionals on function modules) is due to Cunningham and Roy ([33]); our proof is essentially the same as in [15]. Some of the results presented here seem to be new (e.g. prop. 4.8(i)(ii), prop. 4.9, prop. 4.11, th. 4.16, cor. 4.18, prop. 4.19, prop. 4.20, prop. 4.21, prop. 4.22, cor. 4.25, prop. 4.26). Prop. 4.19 and the counter-example on p. 101 have been communicated to us by R. Evans.

It has been shown in [50] and [60] that function modules and bundles of Banach spaces are in one-to-one correspondence and that certain sheaves of Banach spaces define function modules.
A number of important properties of function modules remain valid if def. 4.1 is replaced by a set of weaker axioms (for example: locally compact base space, component spaces which are locally convex but not necessarily normed, etc.). We refer the reader to [60] and the references in [71].
On the other hand, if the component spaces X_k have an additional (algebraic or order-theoretical) structure which is, in a sense, compatible with M-structure, then function modules X in $\prod^{\infty}_{k\in K} X_k$ have also this structure if the respective operations are well-defined (for example, if the X_k are B^*-algebras and $k \mapsto (x(k))^*$ belongs to X for every $x\in X$, then X is also a B^*-algebra).

<u>Supplement: Function module techniques in approximation theory</u>
The reason for the usefulness of function modules in approximation theory is the fact that approximation problems in the function module can be solved locally for the components (see prop. 4.8, in particular 4.8(iv)). As an example of an application of this technique we prove the following Stone-Weierstraß type theorem (for a more general result see [71]):
<u>6.10 Proposition</u>: Let L be a compact Hausdorff space and A a self-adjoint subalgebra of CL with $\underline{1}\in A$. Assume further that X and Y are subspaces of CL such that $Y\subset X$ and $AY\subset Y$ (i.e. Y is an A-module). Then Y is dense in X iff $\{f|_{A_l} \mid f\in Y\}$ is dense in $\{g|_{A_l} \mid g\in X\}$ for every $l\in L$ $(A_l := \{k \mid k\in L,\ f(k) = f(l) \text{ for every } f\in A\})$.

In particular, if A separates the points of L it follows that
Y is dense in X iff $\{f(l) \mid f \in Y\} = \{g(l) \mid g \in X\}$ for every $l \in L$

<u>Proof</u>: By prop. 0.2 there is a compact Hausdorff space K and a continuous function t from L onto K such that $A^{-} = \{h \circ t \mid h \in CK\}$. As in the example on p. 78 we represent CL as a function module with base space K. The assertion then follows immediately from prop. 4.8(iv) (note that Y^{-} is an A^{-}-module). □

Similar techniques have been used by Gierz ([51]) to show that a function module has the approximation property provided that all component spaces have the approximation property.

<u>Supplement: Square Banach spaces</u>

In a sense the simplest function modules are those for which the component spaces are at most one-dimensional. For the case of real scalars these function modules have been discussed in [31] , [77] , [78]. We say that a Banach space X is <u>square</u> if there exists a function module representation of X such that the component spaces are at most one-dimensional. The results of chapter 4, section B, imply that a space is square iff the component spaces in a maximal function module representation are all $\{0\}$ or $\mathbb{K}$.
Further it can be shown that (see [31],[77],[78]):

- square spaces can be characterized by properties of the structure topology of E_X
- every square space is a G-space
- every C_σ-space and every separable G-space is square
- M-spaces which are "not too large" are square
- if X is square and $E:X \to X$ is a contractive projection, then range E is also square.

Square spaces are, in a sense, the M-structure analogues of the L-spaces in L-structure theory since it can be shown that a Banach space is an abstract L-space iff the components in the integral module representation are one-dimensional ([16],[44]). Since L-spaces X are characterized by the property that the commutator of C(X) is just C(X) ([16],[29]) one might expect that a Banach space X is square iff Z(X) and the commutator of Z(X) are identical. The "only if" part is an easy consequence of prop. 4.7 and prop. 4.11. The converse, however, is not true in general as the following counterexample (which is due to R.Evans) shows:

Counterexample: There is a nonsquare Banach space X such that Z(X) is identical with its commutator

Proof: Let K and the X_k be as in the example on p. 130,

$$X := \{(a_1,b_1,c_1,\ldots,(a,b)) \mid a_i \to a, b_i \to b, c_i \to 1/2(a+b)\}.$$

It is clear that X is a function module in $\prod^{\infty}_{k\in K} X_k$. Further, it follows from prop. 4.7 that the operators T which commute with $\{M_h \mid h \in CK\}$ are just the operators

$$T(a_1,b_1,c_1,\ldots,(a,b)) := (\alpha_1 a_1, \beta_1 b_1, \gamma_1 c_1, \ldots, (\alpha a+\gamma b, \delta a+\beta b)),$$

where the $\alpha_i, \beta_i, \gamma_i, \alpha, \beta, \gamma, \delta$ are numbers such that $TX \subset X$. By the definition of X these numbers must satisfy $\gamma = \delta = 0$, $\alpha_i \to \alpha$, $\beta_i \to \beta$, $\gamma_i \to 1/2(\alpha+\beta)$ (apply T for $(1,0,1/2,1,0,1/2,\ldots,(1,0))$,
$(0,1,1/2,0,1,1/2,\ldots,(0,1))$,
and $(1,1,1,\ldots,(1,1))$).

Using this it follows further that $1/2(\alpha+\beta)1/2(a+b) = 1/2(\alpha a+\beta b)$ for $a,b \in \mathbb{K}$ (apply T for $(a,b,1/2(a+b),a,b,1/2(a+b),\ldots,(a,b))$) so that $\alpha = \beta$, i.e. $T\in\{M_h \mid h \in CK\}$.

This proves that Z(X) and its commutator are identical with $\{M_h \mid h \in CK\}$ (so that, in particular, X is not square) □

Note: R. Evans has also shown that, more generally, for every Banach space $\hat{X}$ there is a function module $(K,(X_k)_{k\in K},X)$ such that Z(X) and its commutator are just $\{M_h \mid h \in CK\}$ and $X_k = \hat{X}$ for a suitable $k \in K$.

Remarks concerning chapter 5

Many of the papers which are mentioned in the bibliographical notes of the preceding chapters treat examples where M-structure properties of concrete classes of Banach spaces are determined. We shall omit these references here.

Readers who are primarily interested in ordered Banach spaces are referred to [4], chapter 6 and [89]. M-structure properties of AK-spaces as well as a number of interesting results concerning function module representations of W*-algebras are discussed in [19].

Sections A and B contain the preliminaries from [11] which are necessary to obtain Banach-Stone type theorems for M-finite Banach spaces. We will continue these investigations in chapter 11.

The results at the beginning of section C (th. 5.6-th.5.10) are due to Cunningham, Effros, and Roy ([32]). Our proofs, however, are different. In particular, the proof of th. 5.6 does not depend on a result of Grothendieck as in [32]. Th. 5.13 is new (a preliminary version of 5.13(i) has been considered in [54]).

The theorems in section C show that the centralizer of a dual space X' is completely determined by the L-summands of X. We note that this is also true for the weak*-closed M-ideals of X': it can be shown that every weak*-closed M-ideal is an M-summand (and thus the annihilator of an L-summand in X); [45].

PART II

GENERALIZATIONS OF THE BANACH-STONE THEOREM

Part II: Generalizations of the Banach-Stone theorem

Consider the following problem:

Problem 1: Let X be a Banach space and M and N locally compact Hausdorff spaces such that $C_o(M,X)$ and $C_o(N,X)$ are isometrically isomorphic. Does it follow that M and N are homeomorphic ?

If this is always true we will say that X has the Banach-Stone property (note that the classical Banach-Stone theorem asserts that $\mathbb{K}$ has the Banach-Stone property).

In the next chapters we will apply M-structure methods to investigate problem 1. The main results are proved in chapter 11. For example, it will be shown that

- if X is M-finite, then X has the Banach-Stone property iff the minimum of the M-exponents is one
- if a maximal function module representation of X is known (and this representation is not too pathological) then there is a family of subsets of K_X such that $C_o(M,X) \cong C_o(N,X)$ implies that $\Delta \times M \cong \Delta \times N$ for every Δ in this family. Thus X has the Banach-Stone property if it can be shown that there is a Δ which contains exactly one element.

We will proceed as follows:
Suppose that X is a Banach space for which a maximal function module representation is known and suppose that this is sufficient to determine maximal function module representations of $C_o(M,X)$ and $C_o(N,X)$. If this is the case then corollary 4.17 provides us with the explicit form of every isometric isomorphism I from $C_o(M,X)$ onto $C_o(N,X)$.
Accordingly we will have to discuss the following problems in order to treat problem 1:
Problem 2: For what Banach spaces X is it possible to determine a

maximal function module representation of $C_o(M,X)$ if a maximal function module representation of X is known ?

Problem 3: If X is such a Banach space and $I:C_o(M,X) \to C_o(N,X)$ an isometrical isomorphism, what can be concluded from corollary 4.17 ?

In chapter 7 we will sketch some proofs of the classical Banach-Stone theorem. Chapter 8 contains the precise formulation of problem 1 (Banach-Stone property, strong Banach-Stone property), a number of results of different authors concerning generalizations of the Banach-Stone theorem and a first application of M-structure methods. In chapter 9 we discuss Banach spaces for which the norm topology and the strong operator topology coincide on the centralizer (Banach spaces with a centralizer-norming system). These Banach spaces (more generally, every Banach space with the local cns property) have the property required in problem 2. The proof of this as well as the discussion of some other M-structure properties of $C_o(M,X)$ is the contents of chapter 10. Our main results (which follow from a discussion of problem 3) will be proved in chapter 11. Of particular interest will be the case of M-finite Banach spaces: M-finite Banach spaces with the (strong) Banach-Stone property can be completely characterized. We will see that this result contains all theorems which have been mentioned in chapter 8 (that is, those already obtained without using M-structure methods) as a special case.

Finally, chapter 12 contains bibliographical notes, some remarks and some open problems.

7. The Banach-Stone theorem

The following well-known theorem, usually called the Banach-Stone theorem, will be the starting-point for our investigations in the following chapters. The theorem asserts not only that the topological structure of M is completely determined by the Banach space geometry of C_oM but also that isometric isomorphisms from C_oM onto C_oN must have a particularly simple form (M and N locally compact Hausdorff spaces). We will see that it is dependent on the M-structure properties of the Banach space X whether or not one or both assertions remain valid when C_oM and C_oN are replaced by $C_o(M,X)$ and $C_o(N,X)$.

7.1 Theorem: Let M and N be nonvoid locally compact Hausdorff spaces. If $I:C_oM \to C_oN$ is an isometric isomorphism, then there is a homeomorphism $t:N \to M$ and a continuous map $u:N \to \{\lambda \mid \lambda \in \mathbb{K}, |\lambda| = 1\}$ such that $If = u\cdot(f\circ t)$ (i.e. $(If)(w) = u(w)f(t(w))$) for every $f \in C_oM$, $w \in N$.
In particular, the existence of an isometric isomorphism from C_oM onto C_oN implies that M and N are homeomorphic.

Notes: 1. It is easy to see that, conversely, if t and u are as in the theorem, then $I_{t,u}:C_oM \to C_oN$, $(I_{t,u}f)(w):=u(w)f(t(w))$, is well-defined and an isometric isomorphism.

2. In what follows we shall reserve the letters K,L,... (elements $k,l,k_1,\ldots$) for those topological spaces which are base spaces of function modules under consideration. Other topological spaces, in particular locally compact Hausdorff spaces as in the preceding theorem and its generalizations, will be denoted by M,N,..(elements $v,w,v_1,\ldots$).

The problem in proving theorem 7.1 is twofold. First it has to be shown that there is a family of objects associated with C_oM (functionals, subspaces, subsets,...) which is invariant under isometric isomorphisms and which is indexed by the points of M (this means that M can be reconstructed as a set from the Banach space geometry of C_oM). The second problem is to prove that this correspondence also determines the topology of M.

There are several candidates amongst isometrical invariants having the desired properties. We will sketch proofs which depend on properties of extreme functionals, T-sets, and M-ideals (for detailed proofs we refer the reader to [39], V.8.8 or [81], p.131).

The proof in Banach's book [7] uses differentiability properties of the norm by which it is only possible to treat spaces C_oM for metrizable and compact M. The proof of Stone [83] is similar to our second proof.

<u>First proof of th. 7.1: Extreme functionals</u>

Let $I:C_oM \to C_oN$ be an isometric isomorphism. For $w \in N$ the map $\delta_w \circ I$ ($\delta_w(f) := f(w)$) is an extreme functional on C_oM so that there is a $t(w) \in M$ and a number $u(w) \in \mathbb{K}$, $|u(w)| = 1$, such that $\delta_w \circ I = u(w)\delta_{t(w)}$ (theorem 4.5). Thus, for every $f \in C_oM$ we have $If = u\cdot(f\circ t)$. It is not hard to see that this implies that t is a homeomorphism and that u is continuous.

<u>Second poof of th. 7.1: T-sets</u>

<u>7.2 Definition</u>: Let X be a Banach space. A subset C of X is called a <u>T-set</u> if the norm is additive on C, i.e. $\|x_1+..+x_n\| = \|x_1\|+...+\|x_n\|$ for $x_1,...,x_n \in C$, and C is maximal with respect to this property.

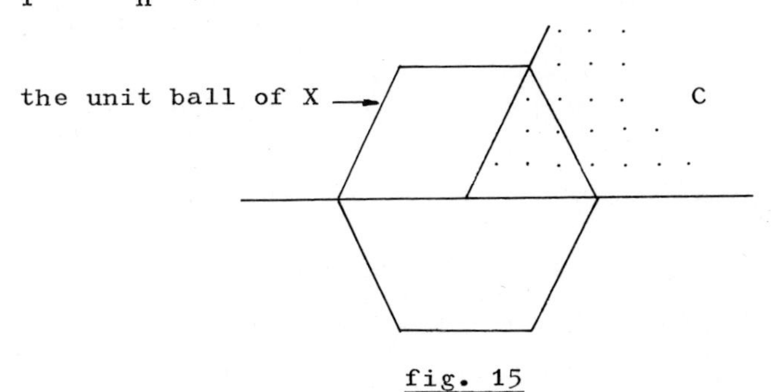

<u>fig. 15</u>

(It is not hard to prove that the T-sets of X are just the cones which are generated by the maximal proper faces of the unit ball of X.)

The T-sets in C_oM are precisely the subsets

$$C_{v,\Delta} := \{f \mid f \in C_oM, |f(v)| = \|f\|, f(v) \in \Delta\} ,$$

where $v \in M$ and Δ is a T-set in $\mathbb{K}$ (this is an easy exercise; a more general result is proved in prop. 8.12 below). Further it is clear that $\lambda \leftrightarrow \Delta_\lambda := \lambda \mathbb{R}^+$ is a one-to-one correspondence between $\{\lambda \mid \lambda \in \mathbb{K}, |\lambda| = 1\}$ and the T-sets of $\mathbb{K}$.

Now suppose that $I: C_oM \to C_oN$ is an isometric isomorphism. I maps T-sets into T-sets so that, for $w \in N$, there is a $\lambda \in \mathbb{K}$ such that $|\lambda| = 1$ and a $v \in M$ such that $I(C_{v,\Delta_\lambda}) = C_{w,\Delta_1}$. With $t(w) := v$ and $u(w) := \lambda^{-1}$ it can be shown that t and u have the claimed properties.

Third proof of th. 7.1: M-ideals

It is clear from the results on p. 36 that the maximal M-ideals of C_oM are just the spaces $J_v := \{f \mid f \in C_oM, f(v) = 0\}$. Thus, if $I: C_oM \to C_oN$ is an isometric isomorphism, there is a $t(w) \in M$ for every $w \in N$ such that $I(J_{t(w)}) = J_w$. It follows that $\ker \delta_{t(w)} \subset \ker(\delta_w \circ I)$ so that $\delta_w \circ I = u(w)\,\delta_{t(w)}$ for a suitable $u(w) \in \mathbb{K}$. We thus have $(If)(w) = u(w) f(t(w))$ for $f \in C_oM$ and $w \in N$ which easily implies that t is a homeomorphism, that u is continuous and that $|u(w)| = 1$ for every $w \in N$.

7.3 Corollary: Let K and L be compact Hausdorff spaces, M and N locally compact Hausdorff spaces. Then $C_o(M,CK)$ and $C_o(N,CL)$ are isometrically isomorphic iff $M \times K$ and $N \times L$ are homeomorphic

Proof: This follows from the fact that $C_o(M,CK) \cong C_o(M \times K)$ and that $C_o(N,CL) \cong C_o(N \times L)$. □

Remark: Thus $C_o(M,CK) \cong C_oN,CK)$ in general does not imply that $M \cong N$ (e.g. $M = \{1\}$, $N = \{1,2\}$, $K = \beta\mathbb{N}$; we have $C_o(M,CK) = C(\{1\},m) \cong m \cong m^2 \cong C(\{1,2\},m) = C_o(N,CK)$ in this case). Cf. problem 3 in chapter 12.

8. The Banach-Stone property and the strong Banach-Stone property

We are now going to investigate the vector-valued generalizations of theorem 7.1. This chapter contains the basic definitions, some examples, a number of theorems by different authors which have been obtained without using M-structure methods, and a first application of the results of part I.

The following lemma shows how homeomorphisms and isometric isomorphisms can be combined to construct isometric isomorphisms between spaces of vector-valued continuous functions:

8.1 Lemma: Let X and Y be Banach spaces, M and N locally compact Hausdorff spaces. Further suppose that $t:N \to M$ is a homeomorphism and that $u:N \to [X,Y]_{iso}$ is a continuous map ($[X,Y]_{iso}$ denotes the set of isometric isomorphisms from X to Y, provided with the strong operator topology).

Then $I_{t,u}:C_o(M,X) \to C_o(N,Y)$,
defined by $(I_{t,u}f)(w) := [u(w)]\, f(t(w))$ (all $f \in C_o(M,X)$, $w \in N$), is an isometric isomorphism

Proof: For $f \in C_o(M,X)$, $w_o \in N$, and $\varepsilon > 0$ choose a neighbourhood W of w_o such that $\|f(t(w)) - f(t(w_o))\| \leq \varepsilon$ and $\|[u(w) - u(w_o)][f(t(w_o))]\| \leq \varepsilon$ for $w \in W$. It follows that

$$\begin{aligned}\|(I_{t,u}f)(w) - (I_{t,u}f)(w_o)\| &= \|u(w)[f(t(w))-f(t(w_o))] + \\ &\quad + [u(w)-u(w_o)][f(t(w_o))]\| \\ &\leq 2\varepsilon\end{aligned}$$

for these w so that $I_{t,u}f$ is continuous at w_o. Since t^{-1} maps compact sets into compact sets and $\|u(w)\| \leq 1$ for every $w \in N$, $I_{t,u}f$ vanishes at infinity, i.e. $I_{t,u}$ is well-defined.

It is clear that $I_{t,u}$ is linear and isometric and it remains to show that $I_{t,u}$ has an inverse.

We first note that $u^{-1}:N \to [Y,X]_{iso}$, $u^{-1}(w) := (u(w))^{-1}$, is continuous (for $w_o \in N$, $y_o \in Y$, and $\varepsilon > 0$ choose $x_o \in X$ such that $u(w_o)x_o = y_o$

and a neighbourhood W of w_o such that $\|u(w)x_o - u(w_o)x_o\| \le \varepsilon$ for $w \in W$; it follows that $\|u^{-1}(w)y_o - u^{-1}(w_o)y_o\| = \|u(w)(u^{-1}(w)y_o - x_o)\|$

$$= \|u(w_o)x_o - u(w)x_o\|$$
$$\le \varepsilon$$

for these w so that $w \mapsto u^{-1}(w)y_o$ is continuous).

By the first part of the proof this implies that $I_{\tilde{t},\tilde{u}}g$ ($\tilde{t} := t^{-1}$, $\tilde{u} := u^{-1} \circ t^{-1}$) is contained in $C_o(M,X)$ for every $g \in C_o(N,Y)$, where $I_{\tilde{t},\tilde{u}}$ is defined similarly to $I_{t,u}$. It is obvious that $I_{\tilde{t},\tilde{u}}$ is an inverse of $I_{t,u}$. □

8.2 Definition: Let X be a Banach space.

(i) We say that X has the Banach-Stone property if the existence of an isometric isomorphism from $C_o(M,X)$ onto $C_o(N,X)$ implies that M and N are homeomorphic
(for every pair M, N of locally compact Hausdorff spaces)

(ii) X is said to have the strong Banach-Stone property if for locally compact Hausdorff spaces the following holds:
for every isometric isomorphism $I: C_o(M,X) \to C_o(N,X)$ there are a homeomorphism $t: N \to M$ and a continuous map $u: N \to [X,X]_{iso}$ such that $(If)(w) = [u(w)][f(t(w))]$ for $f \in C_o(M,X)$ and $w \in N$
(i.e. $I = I_{t,u}$)

Remark: This definition does not agree with the terminology in [23]. What we call "Banach-Stone property" ("strong Banach-Stone property") is there called "weak Banach-Stone property" ("Banach-Stone property").

Examples:

1. Theorem 7.1 may be restated by saying that $\mathbb{K}$ has the strong Banach-Stone property. We recall that it is assumed in all statements that $\mathbb{K}$ is the same scalar field. Thus, if we consider complex Banach spaces, the isometric isomorphisms are assumed to be complex linear. In particular, it does not follow from th. 7.1

that $\mathbb{C}$, as a real Banach space, has the (strong) Banach-Stone property (however, this will be proved later; see th. 8.5).

2. m does not have the Banach-Stone property (see p. 140).
3. l_2^∞ is the simplest space which does not have the Banach-Stone property. By cor. 7.3 it has to be shown that there are non-homeomorphic spaces M, N such that $2M \cong 2N$ (recall that nM is the disjoint union of n copies of M ; $n \in \mathbb{N}$) . In fact, there are compact totally disconnected subsets of the plane with this property (see [56]; other more complicated examples have been constructed in [87], p. 164).
4. Let K be a compact Hausdorff space. By cor. 7.3, CK has the Banach-Stone property iff $K \times M \cong K \times N$ implies that $M \cong N$ (M, N locally compact Hausdorff spaces). $K = \{1\}$ is the only case known to the author where this is true (the examples 2 and 3 are just the cases $K = \beta\mathbb{N}$ and $K = \{1,2\}$, respectively). The case $K = [0,1]$ has been discussed in [49], p. 284: there are non-homeomorphic M, N such that $[0,1]\times M \cong [0,1]\times N$ (M and N are indicated in fig. 16), i.e. C[0,1] fails to have the Banach-Stone property.

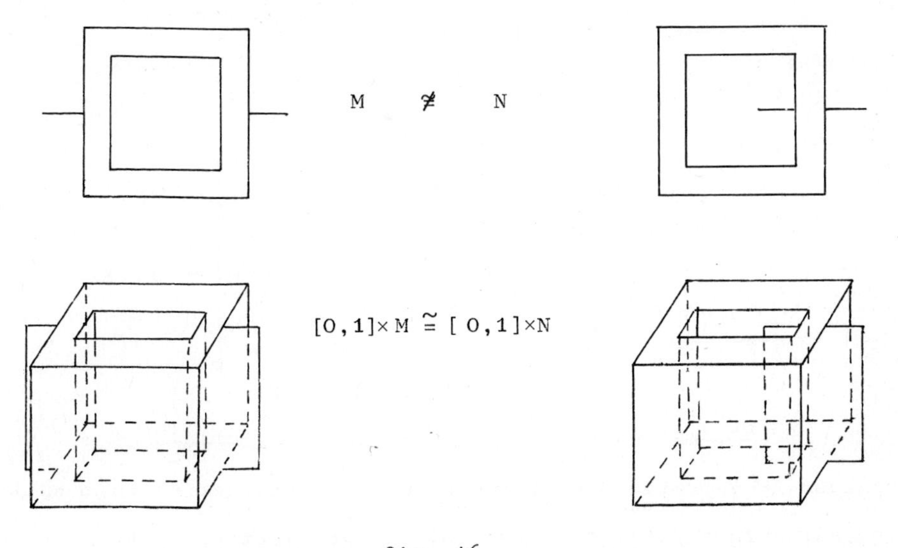

fig. 16

(it is obvious that $[0,1]\times M$ can be deformed into $[0,1]\times N$; the proof of the fact that $M \not\cong N$ is left to the reader).

5. Let $X = \mathbb{R}^3$, provided with the norm $\|(a,b,c)\| := \max\{|a|,(b^2+c^2)^{1/2}\}$ (i.e. $X = \mathbb{R}\times l_2^2$). We will see later that X has the Banach-Stone property but not the strong Banach-Stone property (th. 8.7).

Banach spaces which have the Banach-Stone property or the strong Banach-Stone property have been investigated in a number of papers ([10], [11], [12], [17], [23], [25], [26], [66], [84]). We are going to collect together the most important results which have been obtained without using M-structure methods. Some of these theorems will follow immediately from the fact that the Banach spaces under consideration have a trivial centralizer (th. 8.11). The other results will be derived from our generalizations of the Banach-Stone theorem in chapter 11.

Results of Jerison [66]:

8.3 Definition: Two T-sets C_1, C_2 (cf. def. 7.2) in a Banach space X are called discrepant if either $C_1 \cap C_2 = \{0\}$ or there is a T-set C_3 such that $C_1 \cap C_3 = C_2 \cap C_3 = \{0\}$.

Example:

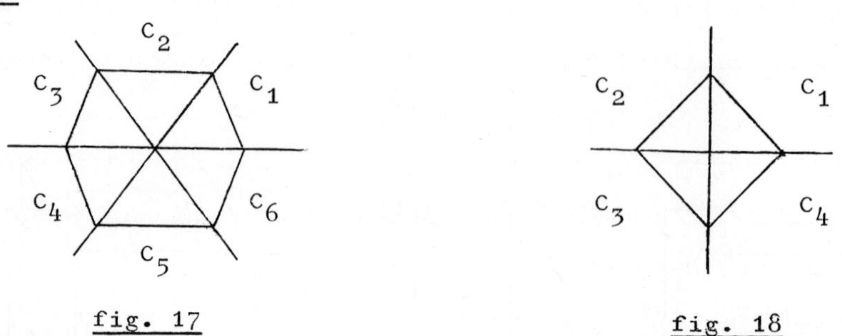

fig. 17 fig. 18

We consider $\mathbb{R}^2$, provided with two norms. The respective unit balls are indicated in fig. 17 and fig. 18. In the first case, every two

T-sets are discrepant whereas in the second only pairs C, -C have this property.

8.4 Theorem [66]: Let X be a real Banach space such that every two T-sets in X are discrepant. Then X has the Banach-Stone property

Proof: see p. 149 below (we will show that X even has the strong Banach-Stone property).

Note: Jerison considers only the case of compact Hausdorff spaces.

8.5 Theorem [66]: Let X be a real strictly convex Banach space, $X \neq \{0\}$. Then X has the strong Banach-Stone property

Proof: see p.148

Results of Cambern ([23],[25],[26])

8.6 Theorem[23],[26]: Let X be a nonzero reflexive space. Then X has the strong Banach-Stone property iff X contains no nontrivial M-summands

Proof: see p.148

Note: The essential idea of Cambern's proof is to show that, if X contains no nontrivial M-summands, the elements of M can be reconstructed from the T-sets of $C_o(M,X)$. Cambern points out in [23] that this is sufficient to guarantee that X has the strong Banach-Stone property.

8.7 Theorem [25]: Let X be a three-dimensional Banach space. Then X has the Banach-Stone property iff X is not isometrically isomorphic to l_3^∞.

In particular, every three-dimensional X such that $X \not\cong l_3^\infty$ which admits non-trivial M-summands has the Banach-Stone property but not the strong Banach-Stone property (see example 5 above).

Proof: see p. 196.

Results of Sundaresan ([84])

8.8 Definition: A Banach space X is called cylindrical if there is a strictly convex smooth space Y such that X is isometrically isomorphic to the product $\mathbb{K}\times Y$ with the supremum norm.

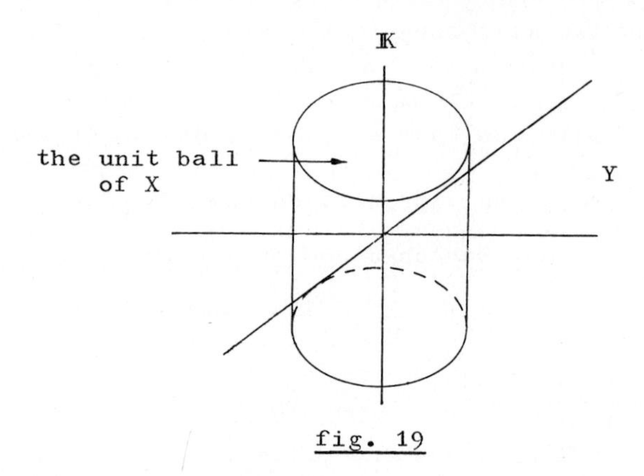

fig. 19

The main result of Sundaresan does not assert that a special class of Banach spaces has the (strong) Banach-Stone property. It is contained here since we will see later that cylindrical spaces which are not two-dimensional have in fact the Banach-Stone property.

8.9 Theorem [84]: Let X be a real cylindrical space which is not two-dimensional, M and N compact Hausdorff spaces which satisfy the first axiom of countability. Suppose that there exists an isometric isomorphism $I: C(M,X) \to C(N,X)$ which satisfies an additional condition (which is too complicated to be described here). Then M and N are homeomorphic.

Proof: see p. 196.

Note: In [84] the theorem is also stated for the two-dimensional case. However, the proof seems to be incorrect for such a space (the first difficulty is that the sets M_1, M_2 in this proof are not well-defined in the two-dimensional case so that there is no canonical way of interpreting the additional condition in the main theorem; moreover it is not clear how lemma 4 of [84] can be used to verify the remark on p. 20, line 15).

In the next chapters we will investigate systematically the problem as to whether a Banach space has the Banach-Stone property (or the strong Banach-Stone property) or not. M-structure methods will play an important role. We will see that the strongest results can be proved for spaces with a "small" centralizer. Generalizations of the Banach-Stone theorem for such spaces are discussed in section B of chapter 11 where they will be obtained as corollaries of more general theorems (for direct proofs see [11]).

As a first application of M-structure we prove that Banach spaces with a "very small" centralizer have the strong Banach-Stone property.

__8.10 Theorem__: Let X and Y be Banach spaces such that Z(X) and Z(Y) are one-dimensional. Further suppose that M and N are nonempty locally compact Hausdorff spaces and that there exists an isometric isomorphism $I:C_o(M,X) \to C_o(N,Y)$.

Then there is a homeomorphism $\tilde{t}:N \to M$ and a continuous map $u:N \to [X,Y]_{iso}$ such that $I = I_{\tilde{t},u}$ (see lemma 8.1).

In particular, the existence of such an isometric isomorphism implies that $X \cong Y$ and that $M \cong N$.

__Proof__: With $K := \beta M$ and $X_k := X$ or $X_k := \{0\}$ if $k \in M$ or $k \in \beta M \setminus M$ we identify $C_o(M,X)$ in a natural way with a function module in $\prod^{\infty}_{k \in K} X_k$ (see p. 78). We claim that $Z(C_o(M,X)) = \{M_h \mid h \in CK\}$.

For $T \in Z(C_o(M,X))$, it follows from prop. 4.11(vi) and the fact that $Z(X) = \mathbb{K}\,Id$ that there is a function $\alpha:\beta M \to \mathbb{K}$ such that $T = M_\alpha$. α is bounded on M by prop. 4.11(i).

α is also continuous at every $v_o \in M$: we choose an $f \in C_o(M,X)$ which is constant and different from zero in a neighbourhood of v_o; since αf is continuous it follows that α must also be continuous on this neighbourhood. Thus there exists an $h \in C(\beta M)$ such that $\alpha|_M = h|_M$, i.e. $T = M_\alpha = M_h$ which proves that $Z(C_o(M,X)) \subset \{M_h \mid h \in CK\}$. Since "$\supset$" is always valid by prop. 4.7(i) we have shown that the natural representation of $C_o(M,X)$ is a maximal function module representation (cf. p. 99).

Similarly we represent $C_o(N,Y)$ as a function module (base space βN, component spaces $Y_l := Y$ for $l \in N$ and $Y_l := \{0\}$ for $l \in \beta N \setminus N$).

By cor. 4.17 there is a homeomorphism $\tilde{t}:\beta N \to \beta M$ and a family of isometric isomorphisms $S_l : X_{\tilde{t}(l)} \to Y_l$ (all $l \in \beta N$) such that $(If)(l) = S_l(f(\tilde{t}(l)))$ (for $f \in C_o(M,X)$, $l \in \beta N$).

For $l \in N$ we have $Y_l = Y \neq \{0\}$ so that $X_{\tilde{t}(l)}$ must also be a nonzero Banach space. Accordingly $\tilde{t}(l) \in M$ so that $\tilde{t}(N) \subset M$. Similarly it follows that $\tilde{t}(N) = M$, i.e. the restriction of $\tilde{t}$ to N (which will also be denoted by $\tilde{t}$) is a homeomorphism from N to M.

We define $u:N \to [X,Y]_{iso}$ by $u(l) := S_l^-$. It is clear from the definition that $I = I_{\tilde{t},u}$, and it remains to show that u is continuous.

For $w_o \in N$ and $x_o \in X$ we choose an $f \in C_o(M,X)$ such that f assumes the constant value x_o in a neighbourhood V of $t(w_o)$. Then $w \mapsto [u(w)](x_o)$ and If coincide on the neighbourhood $t^{-1}(V)$ of w_o, and the latter function is continuous. □

As a corollary we obtain the following

<u>8.11 Theorem</u>: Every Banach space X such that Z(X) is one-dimensional has the strong Banach-Stone property (for examples of such spaces we refer the reader to prop. 5.1). □

This theorem can be applied to prove th. 8.5, th. 8.6, and th. 8.4:

<u>Proof of th. 8.5</u>: This follows from th. 8.11 and cor. 4.23 □

<u>Proof of th. 8.6</u>: Suppose that X is reflexive and that X contains no nontrivial M-summands. Since every M-ideal in X is an M-summand

(prop. 2.2(ii)) it follows that X contains only the trivial M-ideals so that Z(X) must be one-dimensional (cor. 4.23). This proves that X has the strong Banach-Stone property.

Conversely, suppose that X contains a nontrivial M-summand J. For $M = N = \{1,2\}$ we define

$$(If)(w) := \begin{cases} x_2 + x_1^{\perp} & \text{if } w = 1 \\ x_1 + x_2^{\perp} & \text{if } w = 2 \end{cases}$$

(where $f(1) = x_1 + x_1^{\perp}$, $f(2) = x_2 + x_2^{\perp}$; $x_1, x_2 \in J$, $x_1^{\perp}, x_2^{\perp} \in J^{\perp}$).

I is an isometric isomorphism from C(M,X) onto C(N,X), but I is not of the form $I_{t,u}$ (since (If)(w) depends on the values of f at 1 and 2). Consequently X does not have the strong Banach-Stone property. □

<u>Note</u>: The proof of the "only if" part of th. 8.6 did not depend on the reflexivity of X: <u>every</u> Banach space with a nontrivial M-summand fails to have the strong Banach-Stone property.

<u>Proof of th. 8.4</u>: This is a consequence of th. 8.11 and the following proposition (recall that every Banach space can be regarded as a function module with base K_X (th. 4.14) so that, if any two T-sets are discrepant, K_X contains exactly one point (prop. 8.12(iv)), i.e. $Z(X) \cong C(K_X)$ is one-dimensional). □

<u>8.12 Proposition</u>: Suppose that $(K, (X_k)_{k \in K}, X)$ is a function module.

(i) Let k be a point of K such that $X_k \neq \{0\}$ and Δ a T-set in X_k. Then $C_{k,\Delta} := \{ x \mid x \in X,\ \|x\| = \|x(k)\|,\ x(k) \in \Delta \}$ is a T-set in X

(ii) For every T-set C of X there is a $k \in K$ such that $X_k \neq \{0\}$ and a T-set Δ in X_k such that $C = C_{k,\Delta}$

(iii) Let C_{k_1,Δ_1} and C_{k_2,Δ_2} be T-sets as in (i). Then $C_{k_1,\Delta_1} \cap C_{k_2,\Delta_2} = \{0\}$ iff $k_1 = k_2$ and $\Delta_1 \cap \Delta_2 = \{0\}$

(iv) If every two T-sets in X are discrepant, then K contains only one point.

<u>Proof</u>: Before proving the proposition we collect together some general properties of T-sets:

- if C is a T-set, then $C = C^-$
  (since the norm is additive on C^- and $C \subset C^-$)
- every subset on which the norm is additive is contained in a T-set
  (this can be proved by a routine application of Zorn's lemma)
- if C is a T-set, then $C + C \subset C$
  (it is clear from the definition that, for $x,y \in C$, the norm is additive on $\{x+y\} \cup C$ so that $x+y \in C$)
- if C is a T-set and $a \geq 0$, then $aC \subset C$
  ($C+C \subset C$ implies that $nC \subset C$ for $n \in \mathbb{N}$; thus, for $x, x_1, \ldots x_k \in C$ and $n,m \in \mathbb{N}$ we have $\|\frac{n}{m}x + x_1 + \ldots + x_k\| = \frac{1}{m}(\|nx + mx_1 \ldots + mx_k\|) = \frac{1}{m}(\|nx\| + \|mx_1\| + \ldots + \|mx_k\|) = \|\frac{n}{m}x\| + \|x_1\| + \ldots + \|x_k\|$, i.e. the norm is additive on $\{\frac{n}{m}x\} \cup C$; this proves that $\frac{n}{m}C \subset C$ for $n,m \in \mathbb{N}$ so that $aC \in C^- = C$).

(i) It is obvious that the norm is additive on $C_{k,\Delta}$ so that it remains to show that $C_{k,\Delta}$ is maximal with respect to this property. Let x be an element of X such that the norm is additive on $\{x\} \cup C_{k,\Delta}$. Suppose that $\|x(k)\| < \|x\|$. Since $\|x(\cdot)\|$ is upper semicontinuous, there is a neighbourhood U of k such that $\|x(l)\| < 1/2(\|x(k)\| + \|x\|)$ for $l \in U$. $X_k \neq \{0\}$ implies that $\Delta \neq \{0\}$ so that there is an $x_k \in \Delta$ such that $\|x_k\| = 1/2(\|x\| - \|x(k)\|)$ (note that $a\Delta \subset \Delta$ for $a \geq 0$). By lemma 4.2 there exists an $\tilde{x} \in X$ such that $\|\tilde{x}\| = \|x_k\|$, $\tilde{x}(k) = x_k$, $\tilde{x}(l) = 0$ for $l \notin U$ (in particular we have $\tilde{x} \in C_{k,\Delta}$). It follows that $\|x\| = \|x + \tilde{x}\| = \|x\| + \|\tilde{x}\|$, a contradiction.
This proves that $\|x\| = \|x(k)\|$. The norm of X_k is additive on $\{x(k)\} \cup \Delta$ (since, by lemma 4.2, $\Delta = \{x(k) \mid x \in C_{k,\Delta}\}$) so that $x(k) \in \Delta$ and consequently $x \in C_{k,\Delta}$.
(ii) Let T be a T-set in X. For $x \in C$ we define $K_x :=$ $\{k \mid k \in K, \|x\| = \|x(k)\|\}$. The $\|x(\cdot)\|$ are upper semicontinuous so that the K_x are closed and nonempty. For $x_1, \ldots, x_n \in C$ we have $x_1 + \ldots + x_n \in C$ and $\|x_1 + \ldots + x_n\| = \|x_1\| + \ldots + \|x_n\|$. Accordingly $K_{x_1 + \ldots + x_n} \subset$

$K_{x_1} \cap \dots \cap K_{x_n}$ so that $(K_x)_{x \in C}$ has the finite intersection property. Thus there exists a k in $\bigcap \{ K_x \mid x \in C \}$ (note that $X_k \neq \{0\}$ since $C \neq \{0\}$). It is obvious that the norm is additive on $\{x(k) \mid x \in C\}$. Let Δ be a T-set which contains this set. We have $C \subset C_{k,\Delta}$ by definition, and the maximality of C gives $C = C_{k,\Delta}$.

(iii) Suppose that $k_1 \neq k_2$. We choose disjoint neighbourhoods U_1, U_2 of k_1, k_2 and elements x_1, x_2 of X such that $x_1|_{K \setminus U_1} = 0$, $x_2|_{K \setminus U_2} = 0$, $\|x_1\| = \|x_2\| = 1$ (the existence of such vectors follows from lemma 4.2). Then $0 \neq x_1 + x_2 \in C_{k_1,\Delta_1} \cap C_{k_2,\Delta_2}$, i.e. $C_{k_1,\Delta_1} \cap C_{k_2,\Delta_2} = \{0\}$ implies $k_1 = k_2$. It is also clear that $\Delta_1 \cap \Delta_2 = 0$ in this case.

The reverse implication is obvious.

(iv) If K contains more than one point, there are $k_1, k_2 \in K$ such that $k_1 \neq k_2$, $X_{k_1} \neq \{0\} \neq X_{k_2}$. Let Δ_1 and Δ_2 be T-sets in X_{k_1} and X_{k_2}. By (ii) and (iii), the T-sets C_{k_1,Δ_1} and C_{k_2,Δ_2} are not discrepant. □

Finally it should be noted that the class of spaces X for which Z(X) is one-dimensional is strictly larger than the classes of spaces for which the theorems 8.4, 8.5, 8.6 apply: l^1 is a non-reflexive spaces which is not strictly convex and for which only pairs C, -C of T-sets are discrepant; however, by prop. 5.1(vi) we have $Z(l^1) = \mathbb{K}\,\mathrm{Id}$.

9. Centralizer-norming systems

In this chapter we discuss a property of certain Banach spaces X which (as we will see in chapter 10) guarantees that M-structure properties of $C_o(M,X)$ (M a locally compact Hausdorff space) can be derived from M-structure properties of X. In view of the generalizations of the Banach-Stone theorem the centralizer of $C_o(M,X)$ will be of particular interest. It is an easy consequence of the results in part I that, for $T \in Z(C_o(M,X))$, there is a family $(T_v)_{v\in M}$ of operators in Z(X) such that $(Tf)(v) = T_v(f(v))$ (for details see chapter 10). It is not hard to prove that $v \mapsto T_v$ is a continuous mapping from M into Z(X) with the strong operator topology. However, it is often necessary to know that this mapping is in fact continuous when Z(X) is provided with the norm topology.

Thus the following problem arises:

> Let X be a Banach space. Under what conditions on X is it possible to guarantee that
>
> (*) every continuous mapping from a locally compact Hausdorff space into Z(X) with the strong operator topology is in fact continuous when Z(X) is provided with the norm topology ?

A sufficient condition is easily established: (*) is trivially true for Banach spaces X for which the norm topology and the strong operator topology coincide on Z(X). Banach spaces with this property will be very important for the following considerations:

9.1 Definition: Let X be a Banach space. A finite family $x_1,\ldots,x_n$ in X is called a centralizer-norming system (abbr.: cns) if there is a number $r > 0$ such that $r\|T\| \leq \max\{\|Tx_i\| \mid i=1,\ldots,n\}$ for every $T \in Z(X)$.

It is obvious that X has a cns iff the strong operator topology and the norm topology coincide on Z(X).

The following proposition asserts that the Banach spaces which have a cns are precisely the spaces for which a modified form of (*) is valid:

9.2 Proposition: For a Banach space X the following are equivalent:

a) X has a cns

b) every continuous map from a topological space into Z(X) with the strong operator topology is continuous when Z(X) is provided with the norm topology

Proof:

"a ⇒ b": This is obvious.

"b ⇒ a": Suppose that X has no cns. By $P_{fin}(X)$ we denote the family of nonvoid finite subsets of X, and M means the set $P_{fin}(X) \cup \{\infty\}$. We will say that a subset U of M is a neighbourhood of $\Delta \in P_{fin}(X)$ (of ∞) if $\Delta \in U$ (if there is a $\Delta_o \in P_{fin}(X)$ such that $\{\Delta | \Delta \in P_{fin}(X), \Delta \supset \Delta_o\} \cup \{\infty\} \subset U$). Let $\mathfrak{T}$ be the topology associated with this definition of "neighbourhood" (i.e. $O \in \mathfrak{T}$ iff for every $v \in O$ there is a neighbourhood U of v such that $U \subset O$).

We note that $\{\Delta | \Delta \in P_{fin}(X), \Delta \supset \Delta_o\} \cup \{\infty\}$ is an open neighbourhood of ∞ for every $\Delta_o \in P_{fin}(X)$ and that $(M, \mathfrak{T})$ is a Hausdorff space.

Since X has no cns there exists an operator $T_\Delta \in Z(X)$ for every $\Delta \in P_{fin}(X)$ such that $\max\{\|T_\Delta x\| \mid x \in \Delta\} < (1/\text{card } \Delta)\|T_\Delta\|$ (without loss of generality we may assume that $\|T_\Delta\| = 1$).

We define $\omega : M \to Z(X)$ by $\omega(\Delta) := T_\Delta$ for $\Delta \in P_{fin}(X)$ and $\omega(\infty) := 0$. It is clear from the definitions that ω is continuous when Z(X) is provided with the strong operator topology. However, ω is not continuous with respect to the norm topology on Z(X) since ∞ has no neighbourhood such that $\|\omega(v)\| \leq 1/2$ for v in this neighbourhood. This contradiction proves that b implies a. □

Note: It is easy to see that $P_{fin}(X)$ can be replaced by $P_{fin}(Y)$, where Y is a dense subset of X. It follows that M can be chosen to be $\alpha\mathbb{N}$ if X is separable (this has been noted by R. Evans).

Examples:

1. Every M-finite Banach space (cf. the definition on p. 110) has a cns. This follows from the fact that on a finite-dimensional space any two locally convex Hausdorff topologies are equivalent.
   Note: Moreover it can be shown that for every M-finite Banach space X there exists an $x \in X$ such that $\|T\| = \|Tx\|$ for every $T \in Z(X)$:
   if X is M-finite and $X = \prod_{i=1}^{n}{}^{\infty} X_i$ a representation of X as in prop. 5.4(v), we choose $x_i \in X_i$ such that $\|x_i\| = 1$ for every i and define $x := (x_1,\dots,x_n)$; it is clear from prop. 5.4(iv) that $\|T\| = \|Tx\|$ for every $T \in Z(X)$.
2. Let L be a nonvoid locally compact Hausdorff space. Then C_oL has a cns iff L is compact:

   In the compact case, $\{\underline{1}\}$ is a cns (and similarly as in example 1, $\|T\| = \|T\underline{1}\|$ for every $T \in Z(C_oL)$); this follows from prop. 5.14. If L is not compact and $f_1,\dots,f_n \in C_oL$ are arbitrary, we choose for $\varepsilon > 0$ a nonvoid open set U such that $\max_i \|f_i|_U\| < \varepsilon$. But then we have $\|M_h\|\varepsilon > \max_i \|M_h f_i\|$ for a suitable function $h \in C^bL$ such that $\|h\| = 1$ and $h|_{L\setminus U} = 0$, i.e. $f_1,\dots,f_n$ is not a cns for C_oL.
3. Let A be a B^*-algebra with unit e. Then $\|T\| = \|Te\|$ for every $T \in Z(A)$ (this follows from prop. 5.15) so that $\{e\}$ is a cns for A.
   Similarly $\{e\}$ is a cns for the real Banach space A_{sa}.
4. Let K be a nonempty compact convex set in a locally convex Hausdorff space and AK the space of $\mathbb{K}$-valued continuous affine functions on K. We provide AK with the supremum norm. It is well-known that $E_{AK} \subset \{\lambda\delta_k \mid k \in K, \lambda \in \mathbb{K}, |\lambda| = 1\}$.
   This implies that $\|T\| = \|T\underline{1}\|$ for every $T \in Z(AK)$ (since $\|T\| = \sup\{|a_T(p)| \mid p \in E_{AK}\} = \sup\{|p(T\underline{1})| \mid p \in E_{AK}\} = \|T\underline{1}\|$), i.e. $\{\underline{1}\}$ is a cns in AK.

5. For every dual space X' there exists a $p \in X'$ such that $\|Tp\| = \|T\|$ for every $T \in Z(X')$, i.e. $\{p\}$ is a cns (choose $p \in X'$ as in th. 5.13(ii); in a maximal function module representation of X' the $T \in Z(X')$ are of the form M_h so that $\|Tp\| = \|T\|$ follows from $\|\rho(p)(\cdot)\| = 1$).

The Banach spaces X which we considered in the preceding examples have the property that, if a cns exists, there is in fact an element $x \in X$ such that $\|Tx\| = \|T\|$ for every $T \in Z(X)$. This is not true in general as the following examples show.

6. For $\mathbb{K} = \mathbb{R}$ we consider the space X_s of example 2 on p.100. Since $Z(X_s) = \{M_h \mid h \in C([0,2\pi]),\ h(0) = h(2\pi)\}$ it follows that a finite family $f_1,\dots,f_n \in X_s$ is a cns iff $\min_t \max_i |f_i(t)| > 0$.
   a) If $0 < s \le 1$, then $\{f_s\}$ (where $f_s(t) := 1 - t(1-s)$) is a cns for X_s: we have $\|Tf_s\| \ge s\|T\|$ for every $T \in Z(X_s)$.
   Also it is easy to see that for $f_1,\dots,f_n \in X_s$ such that $\|f_1\| = \dots = \|f_n\| = 1$ and $\max_i \|Tf_i\| \ge r\|T\|$ (all $T \in Z(X_s)$) necessarily $r \le s$ holds. Thus it is possible that r is of necessity small in def. 9.1 (provided that the vectors in the cns are assumed to be normalized).
   b) If $-1 \le s < 0$ then every $f \in X_s$ vanishes at a point of $[0,2\pi]$ so that there is no cns containing only one element. However, it is easy to see that there are functions $f_1, f_2 \in X_s$ such that $\min_t \max_i |f_i(t)| > 0$ so that f_1, f_2 is a cns in X_s.
7. Example 6b depends essentially on the well-known fact from elementary calculus that a continuous function from an interval to $\mathbb{R}$ which changes sign must vanish at some point. This result can be thought of as a special case of the Borsuk-Ulam theorem from algebraic topology (see, for example, [1], p.485). Using this theorem it is possible to show that, for $n \in \mathbb{N}$, there are spaces which have a cns consisting of n+1 elements but no cns consisting

of n elements.

Let $n \in \mathbb{N}$ and X the space of example 3 on p. 101.

$Z(X) = \{M_h \mid h \in C(S_n),\ h(s) = h(-s) \text{ for every } s \in S_n\}$ implies that $f_1,..,f_k \in X$ is a cns iff $\min_s \max_i |f_i(s)| > 0$, i.e. iff there is no point $s_o \in S_n$ such that $f_1(s_o) = \ldots = f_k(s_o) = 0$. The Borsuk-Ulam theorem says precisely that such an s_o always exists if $k \leq n$ (in the case of real scalars; for $\mathbb{K} = \mathbb{C}$ we apply the theorem to the real and imaginary parts of the functions under consideration so that there is an s_o such that $f_1(s_o) = \ldots = f_k(s_o) = 0$ if $2k \leq n$). Thus X has no cns consisting of n elements (in the complex case: of $[n/2]$ elements). However, it is not difficult to find a cns containing n+1 elements (for example the functions $g_1,\ldots,g_{n+1}$ are a cns, where g_i denotes the projection onto the i'th component).

By the following proposition it is possible to find further examples of spaces with a cns:

<u>9.3 Proposition</u>:

(i) Let $(X_i)_{i \in I}$ be a family of Banach spaces and suppose that there are $x_1^i,\ldots,x_{n_i}^i \in X_i$, $r_i > 0$ such that $\max\{\|Tx_j^i\| \mid j=1,..n_i\} \geq r_i\|T\|$ for $T \in Z(X_i)$ (all $i \in I$).

Then $X := \prod_{i \in I}^{\infty} X_i$ has a cns provided that $n := \max_i n_i < \infty$, $r := \inf_i r_i > 0$, $\sup\{\|x_j^i\| \mid i \in I,\ j=1,\ldots,n_i\} < \infty$. In particular every finite product has a cns if all components have this property.

Conversely, if $\prod_{i \in I}^{\infty} X_i$ has a cns, then all components also have a cns.

(ii) Let K be a nonvoid compact Hausdorff space, X a Banach space. Then $C(K,X)$ has a cns iff X has a cns.

(iii) Every M-summand (but not necessarily every M-ideal) of a Banach space with a cns also has a cns.

Note: It follows from the results in [13], th.4.5, that $X\hat{\otimes}_{\varepsilon}Y$ has a cns if X and Y have a cns (note that this result contains (ii) as a special case since $C(K,X) \cong CK\hat{\otimes}_{\varepsilon}X$).

Proof:

(i) Without loss of generality we may assume that $n_i = n$ for every $i \in I$. Let $x_j := (x_j^i)_{i\in I}$ for $j = 1,\dots,n$. We claim that $r\|T\| \le \max_j \|Tx_j\|$ for every $T \in Z(X)$.

We have $T \in Z(X)$ iff $T = \prod_{i\in I} T_i$, where $T_i \in Z(X_i)$ for every $i \in I$ and $\sup_i \|T_i\| < \infty$ (the proof of this fact which is similar to the proof on p. 18 in chapter 1 is left to the reader). Thus

$$\begin{aligned}\max_j \|Tx_j\| &= \max_j \|(T_i x_j^i)_{i\in I}\| \\ &= \max_j \sup_i \|T_i x_i^j\| \\ &= \sup_i \max_j \|T_i x_i^j\| \\ &\ge r \sup_i \|T_i\| \\ &= r\|T\| \qquad \text{for every } T \in Z(X).\end{aligned}$$

Conversely, if $x_j = (x_j^i)_{i\in I}$, $j=1,\dots,n$ is a cns for X and $i_o \in I$, then $x_1^{i_o},\dots,x_n^{i_o}$ is a cns in X_{i_o}: for $T_{i_o} \in Z(X_{i_o})$ we consider $T = \prod_{i\in I} \tilde{T}_i$, where $\tilde{T}_i := T_{i_o}$ if $i=i_o$ and $\tilde{T}_i := 0$ if $i \ne i_o$. We have $T \in Z(X)$ so that $\max_j \|T_{i_o} x_j^{i_o}\| = \max_j \|Tx_j\| \ge r\|T\| = r\|T_{i_o}\|$.

(ii) Suppose that $\max\{\|Tx_i\| \mid i=1,\dots,n\} \ge r\|T\|$ for every $T \in Z(X)$. We claim that $\underline{x}_1,\dots,\underline{x}_n$ ($\underline{x}_i$ denotes the function which assumes the value x_i at every point) is a cns for $C(K,X)$.

For $T \in Z(C(K,X))$ it follows from prop. 4.7(iv) that $T = \prod T_k$, where $T_k \in Z(X)$ for every $k \in K$ and $\sup_k \|T_k\| < \infty$. Accordingly

$$\begin{aligned}\max_i \|T\underline{x}_i\| &= \max_i \max_k \|T_k x_i\| \\ &\ge r \sup_k \|T_k\| \\ &= r\|T\|.\end{aligned}$$

Conversely, suppose that $\max\{\|Tf_i\| \mid i=1,\dots,n\} \ge r\|T\|$ for every T in $Z(C(K,X))$, i.e. that $f_1,\dots,f_n$ is a cns for $C(K,X)$. We claim that $f_1(k_o),\dots,f_n(k_o)$ is a cns in X for every k_o in K.

Suppose that $k_o \in K$ and that there is a $T_o \in Z(X)$ such that $\max_i \|T_o(f_i(k_o))\| < r\|T_o\|$. It follows that $\max_i \|(h \otimes T_o)f_i\| < r\|h \otimes T_o\|$ for a suitable $h \in CK$ ($((h \otimes T_o)f)(k) := h(k)T_o(f(k))$). We have $h \otimes T_o \in Z(C(K,X))$ (this follows at once from 4.7(iv)) in contradiction to the fact that $f_1,\dots,f_n$ is a cns in $C(K,X)$.

(iii) The assertion concerning M-summands is a special case of (i) since $X \cong J \times J^{\perp}$ (the product provided with the supremum norm) for every M-summand J of X.

The M-ideal c_o of c has no cns although $(1,1,\dots)$ is a cns for c (cf. example 2 on p. 154). □

We are now going to investigate centralizer-norming systems by means of a maximal function module representation of X:

<u>9.4 Proposition</u>: Let $(K,(X_k)_{k\in K},X)$ be a function module such that $Z(X) = \{M_h \mid h \in CK\}$ (recall that by the results of chapter 4 every Banach space can be put in this form).

For $x_1,\dots,x_n \in X$ and $r > 0$ the following are equivalent:

a) $\max_i \|Tx_i\| \geq r\|T\|$ for every $T \in Z(X)$

b) $\max_i \|x_i(k)\| \geq r$ for every $k \in K$

(i.e. $x_1,\dots,x_n$ is a cns iff $\inf_k \max_i \|x_i(k)\| > 0$)

<u>Proof</u>:

<u>"a ⇒ b"</u>: Assume that there is a $k \in K$ such that $\max_i \|x_i(k)\| < r$. Choose a neighbourhood U of k such that $\sup\{\|x_i(l)\| \mid i=1,\dots,n,\ l \in U\} < r$ and $h \in CK$ such that $\|h\| = h(k) = 1$ and $h|_{K \setminus U} = 0$.

We then have $M_h \in Z(X)$ and $\max_i \|M_h x_i\| < r\|M_h\|$, a contradiction.

<u>"b ⇒ a"</u>: This is an immediate consequence of of $Z(X) = \{M_h \mid h \in CK\}$ □

<u>9.5 Corollary</u>: Let X be a Banach space, $x_1,\dots,x_n \in X$, $r > 0$.

Suppose that $\max_i |p(x_i)| \geq r$ for every $p \in E_X$. Then $\max_i \|Tx_i\| \geq r\|T\|$ for every $T \in Z(X)$, i.e. $x_1,\dots x_n$ is a cns

<u>Proof</u>: This follows from prop. 9.4 and the fact that, for $k \in K$ such

that $X_k \neq \{0\}$, $\max_i \|x_i(k)\| = \max_i |p(x_i)| \geq r$ for a suitable $p \in E_X$ (th. 4.5); these k are dense in K and the $\|x_i(\cdot)\|$ are upper semi-continuous so that $\max_i \|x_i(k)\| \geq r$ for every $k \in K$.

(<u>Alternative proof</u>:
$$\begin{aligned} r\|T\| &= r\|a_T\| \\ &= r \sup\{ |a_T(p)| \mid p \in E_X\} \\ &\leq \sup\{| a_T(p)p(x_i) | \mid p \in E_X, i=1,\dots,n\} \\ &= \sup\{| p(Tx_i) | \mid p \in E_X, i=1,\dots,n\} \\ &= \max_i \|Tx_i\| \end{aligned} \quad ;$$
this shows that in this case the strong operator topology and the norm topology coincide on Mult(X) and not only on Z(X).) □

<u>Note</u>: This corollary can be used to prove that $\{\underline{1}\}$ is a cns for spaces CK and AK (cf. examples 2 and 4).

<u>9.6 Corollary</u>: If X has a cns, then in any function module representation of X all component spaces X_k are nonzero. In particular this is true for maximal function module representations so that $K_X = K_X^*$ (cf. the definition of K_X^* on p. 103).

<u>Proof</u>: For maximal function module representations this is clear from prop. 9.4, the general case follows from the results of chapter 4, section B (the assertion can also be proved directly as in the proof of prop. 9.4,a $\Rightarrow$ b). □

Let X be a Banach space and K_X a compact Hausdorff space such that $Z(X) \cong C(K_X)$. For the rest of this chapter we will assume that X has been identified with a function module in $\prod^{\infty}_{k \in K_X} X_k$ as in th. 4.14 (so that, in particular, $Z(X) = \{M_h \mid h \in CK_X\}$).

Our investigations in chapter 10 will show that in order to discuss a maximal function module representation of $C_o(M,X)$ (M a locally compact Hausdorff space) it is not necessary to assume that X has a cns but that it is sufficient to know that a weaker condition is satisfied (see def. 9.8).

<u>9.7 Definition</u>: Suppose that $k \in K_X$. A finite family $x_1,\dots,x_n \in X$ is called a <u>local centralizer-norming system</u> (abbr.: <u>local cns</u>) <u>for k</u> if there is a neighbourhood U of k such that $\inf_{l \in U} \max_i \|x_i(l)\| > 0$.

<u>Remarks/Examples</u>:

1. It is clear from the compactness of K_X and prop.9.4 that X has a cns iff every $k \in K_X$ has a local cns.
2. It is obvious that $k \in K_X^*$ if k admits a local cns (moreover, $\{k \mid k \in K_X$, k admits a local cns$\}$ is an open subset of K_X^*). We will see later that the converse is not true (counterexample 2 on p. 164).
3. As "$k \in K_X^*$" the assertion "$x_1,\dots,x_n$ is a local cns for k" does not only depend on k but also on the representation of X as a function module with base space K_X. This will cause no confusion if we restrict our investigation to a fixed representation.
4. Consider $X = C_oL$, where L is a locally compact non-compact Hausdorff space. We represent X as a function module with base space $K_X = \beta L$ as in prop. 5.14. Then every $k \in K_X^*$ ($= L$) has a local cns (every function f which does not vanish at k is a local cns for k).

We will see that if a Banach space X has the property of C_oL in the preceding example it is possible to determine a maximal function module representation of $C_o(M,X)$ (M a locally compact Hausdorff space) whenever such a representation of X is known, a fact which will of particular importance in generalizing theorem 7.1. Fortunately it seems that only very pathological spaces fail to have this property, a precise formulation of which is contained in the following definition:

<u>9.8 Definition</u>: Let X be as above. We say that X has the <u>local cns property</u> if every $k \in K_X^*$ has a local cns (i.e. for every $k \in K_X$ such that $X_k \neq \{0\}$ there are elements in X which do not behave too

pathologically on a suitable neighbourhood of k).
We note that, since maximal function module representations are equivalent, this property does not depend on the particular function module representation of X with base space K_X.

Remarks:

1. We have already noted that it is not obvious that there are spaces which do not have the local cns property. Examples will be given on p. 163.
2. K_X^* is necessarily an open subset of K_X if X has the local cns property.

The following proposition guarantees that if neither the component space X_{k_o} nor the neighbourhood filter of a point $k_o \in K_X^*$ is too complicated, then k_o has a local cns. This will sometimes be useful in deciding whether or not a space has the local cns property. Also this proposition indicates how to go about constructing Banach spaces which fail to have the local cns property.

9.9 Proposition: Let X be as above and $k_o \in K_X^*$. Further suppose that there is a sequence $\tilde{x}_1, \tilde{x}_2, \ldots$ in X such that $\overline{\text{lin}}\{h\tilde{x}_i \mid h \in CK_X, i \in \mathbb{N}\} = X$. Each of the following conditions implies that k_o has a local cns:

(i) k_o is an interior point of K_X^*

(ii) if $O_1, O_2, \ldots$ is a sequence of open subsets in K_X such that $O_1 \supset O_2 \supset \ldots$ and $k_o \in O_n^-$ for every $n \in \mathbb{N}$, then there exists a sequence $k_1, k_2, \ldots$ in K_X such that $k_n \in O_n \cap K_X^*$ for every n and $\lim k_n = k_o$

(this is true, for example, if there exists a countable basis of the neighbourhood system of k_o)

Proof: We will prove that if (i) or (ii) is satisfied and $\tilde{x}_1, \ldots, \tilde{x}_n$ is not a local cns for k_o for every $n \in \mathbb{N}$, then there is a function $\alpha : K_X \to \mathbb{K}$ such that $M_\alpha X \subset X$ (i.e. $M_\alpha \in Z(X)$) but $M_\alpha \notin \{M_h \mid h \in CK_X\}$ in contradiction to the fact that the function module representation of X is maximal. For simplicity we will assume that k_o is not an isolated point of K_X (in this case k_o has obviously a local cns).

(i) Let U_o be an open neighbourhood of k_o such that $X_k \neq \{0\}$ for every $k \in U_o^-$ and suppose that $\tilde{x}_1, \ldots, \tilde{x}_n$ is not a local cns for k_o for every $n \in \mathbb{N}$.

In particular $\{\tilde{x}_1\}$ is not a local cns so that there is a $k_1 \in U_o$ such that $\|\tilde{x}_1(k_1)\| < 1$ (we may assume that $k_1 \neq k_o$ since $\|\tilde{x}_1(\cdot)\|$ is upper semicontinuous and k_o is not isolated). Let U_1 be a closed neighbourhood of k_1 such that $k_o \notin U_1$ and $\|\tilde{x}_1(l)\| < 1$ for $l \in U_1$.

Since $\{\tilde{x}_1, \tilde{x}_2\}$ is not a cns for k_o there is a k_2 ($\neq k_o$) such that $\max\{\|\tilde{x}_1(k_2)\|, \|\tilde{x}_2(k_2)\|\} < 1/2$, $k_2 \in U_o \setminus U_1$. Let U_2 be a closed neighbourhood of k_2 contained in $U_o \setminus U_1$ such that $k_o \notin U_2$ and $\max\{\|\tilde{x}_1(l)\|, \|\tilde{x}_2(l)\|\} < 1/2$ for $l \in U_2$.

In this way we obtain a sequence $k_1, k_2, \ldots$ of points in U_o and neighbourhoods U_n of k_n such that

- $k_o \notin U_n$
- $U_{n+1} \subset U_o \setminus (U_1 \cup \ldots \cup U_n)$
- $\sup\{\|\tilde{x}_i(l)\| \mid i=1,\ldots,n,\ l \in U_n\} < 1/n$ (all $n \in \mathbb{N}$).

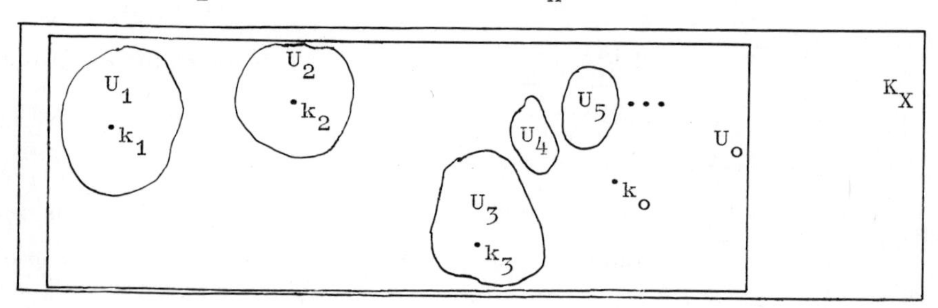

fig. 20

We choose functions $h_n \in CK_X$ such that $\|h_n\| = h_n(k_n) = 1$, $h_n|_{K_X \setminus U_n^o} = 0$ and define $\alpha: K_X \to \mathbb{K}$ by $\alpha(k) := \sum_n h_n(k)$ (note that for every k at most one term in this series is nonzero).

We claim that $M_\alpha X \subset X$ and that $M_\alpha \notin \{M_h \mid h \in CK_X\}$.

Since $\|(\sum_{m\geq n} h_m)\tilde{x}_{n_o}\| \leq 1/n$ for $n \geq n_o$ we have $M_\alpha \tilde{x}_{n_o} = \lim_n (\sum_{m=1}^{n} h_m)\tilde{x}_{n_o}$ so that (since $\sum_{m=1}^{n} h_m \in CK_X$ and X is a function module) $M_\alpha \tilde{x}_{n_o} \in X^- = X$. This implies that $M_\alpha X \subset X$ (since the $h\tilde{x}_n$ span X; $h \in CK_X$, $n \in \mathbb{N}$).

Suppose that $M_\alpha = M_h$ for an $h \in CK_X$. We have $\alpha|_{U_o^-} = h|_{U_o^-}$ since $U_o^- \subset K_X^*$, but α is not continuous on U_o^-: Since $\{k_1, k_2, \ldots\}$ is not closed in K_X (this set is obviously not compact) there is a $k^* \in \{k_1, \ldots\}^- \setminus \{k_1, \ldots\}$; we have $k^* \notin U_1^o \cup U_2^o \cup \ldots$, $k^* \in U_o^-$ so that $\alpha(k^*) = 0$, but every neighbourhood of k^* contains elements k_n (n sufficiently large) for which $\alpha(k_n) = 1$.

(ii) Suppose that $\tilde{x}_1, \ldots, \tilde{x}_n$ is not a local cns for k_o (all $n \in \mathbb{N}$). $X_{k_o} \neq \{0\}$ implies that there is an index n_o such that $\tilde{x}_{n_o}(k_o) \neq 0$. Without loss of generality we may assume that $n_o = 1$ and that $\|\tilde{x}_1(k_o)\| \geq 1$. For $n \in \mathbb{N}$ we define

$$U_n := \{k \mid \max\{\|\tilde{x}_i(k)\| \mid i=1,\ldots,n\} < 1/n\} .$$

Since $\tilde{x}_1, \ldots, \tilde{x}_n$ is not a local cns for k_o we have $k_o \in U_n^-$. The U_n are open and decreasing so that there exists a sequence $(k_n)_{n\in\mathbb{N}}$ such that $k_n \in U_n \cap K_X^*$ and $\lim k_n = k_o$.
$\|\tilde{x}_1(k_o)\| \geq 1$ implies that $k_n \neq k_o$ for every $n \in \mathbb{N}$.
We construct a subsequence $k_{n_1}, k_{n_2}, \ldots$ of $k_1, k_2, \ldots$ and neighbourhoods V_{n_i} of k_{n_i} such that $V_{n_i} \cap V_{n_j} = \emptyset$ for $i \neq j$. Then, similarly to the proof of (i), we define $\alpha := \sum_i h_i$, where $h_i \in CK_X$ is a function such that $h_i(k_{n_i}) = \|h_i\| = 1$, $h_i|_{K_X \setminus V_{n_i}^o} = 0$. As in (i) it can be shown that $M_\alpha X \subset X$ (so that $M_\alpha \in Z(X)$), but there is no function h in CK_X such that $M_\alpha = M_h$: if $M_\alpha = M_h$, then $\alpha(k_n) = h(k_n)$ for n=0,1,...; but $h(k_o) = \alpha(k_o) = 0 \neq 1 = \lim_i h(k_{n_i})$. □

There are Banach spaces which do not have the local cns property:

<u>Counterexample 1</u>: There is a Banach space X such that all component spaces are at most one-dimensional (i.e. X is square; as in the preceding proposition we regard

X as a function module with base space K_X) but X does not have the local cns property

<u>Counterexample 2</u>: There is a Banach space X which does not have the local cns property and for which $K_X = \alpha\mathbb{N}$

<u>Sketch of proof</u>:

1. Let A be the disjoint union of countably many copies of [0,1]. A^* denotes the quotient of A where the zeros in all copies of [0,1] have been merged together into a single point 0^*:

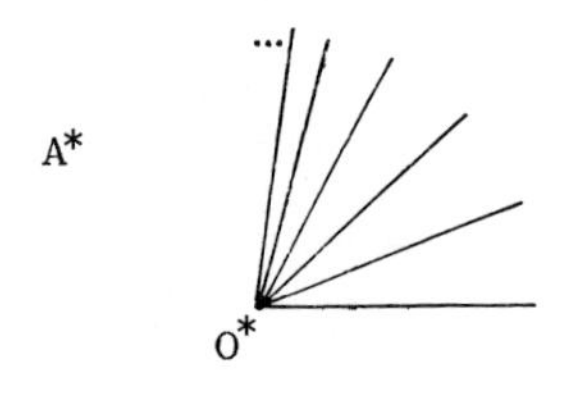

<u>fig. 21</u>

A^* is completely regular, and we define K to be the Stone-Čech compactification βA^* of A^*.

Let $f:K \to \mathbb{R}$ be the function

$$f(k) := \begin{cases} 1/n & \text{if } k \text{ is in the n'th copy of } [0,1] \text{ and } k \neq 0^* \\ 1 & \text{if } k = 0^* \\ 0 & \text{if } k \in \beta A^* \setminus A^* \end{cases} .$$

f is upper semicontinuous on K so that $X := \{hf \mid h \in CK\}^-$ is a function module in $\prod^{\infty}_{k\in K} X_k$ (where $X_k = \mathbb{K}$ or $X_k = \{0\}$ if $k \in A^*$ or $k \in \beta A^* \setminus A^*$). Every function $\alpha:K \to \mathbb{K}$ such that $M_\alpha X \subset X$ must be continuous and bounded on A^* which easily implies that $Z(X) = \{M_h \mid h \in CK\}$, i.e. $K = K_X$, and the identical representation of X is maximal. 0^* is a point of K_X^* ($= A^*$) which does not have a local cns so that X fails to have the local cns property.

2. Let $\mathfrak{U}$ be a free ultrafilter on $\mathbb{N}$. We define $K := \alpha\mathbb{N}$ and $X_k := \mathbb{K}$ for $k \in \mathbb{N}$ and $X_\infty := \prod^1_{U\in\mathfrak{U}} \mathbb{K}_U$ (where $\mathbb{K}_U := \mathbb{K}$ for every $U \in \mathfrak{U}$).

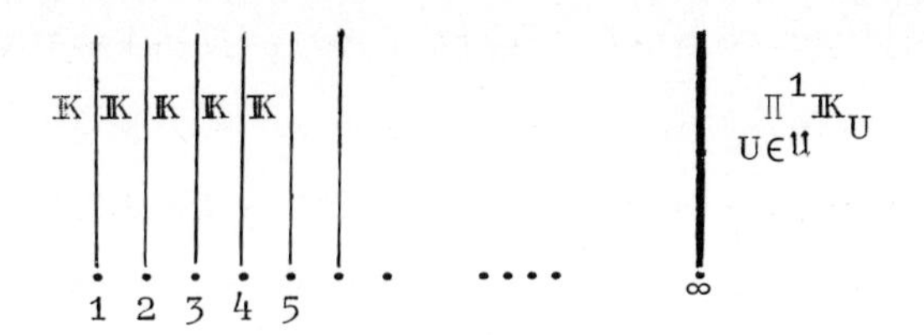

fig. 22

Further, for $U \in \mathfrak{U}$, let $x_U \in \prod_{k\in K}^{\infty} X_k$ be defined by

$$x_U(k) := \begin{cases} 0 & k \in U \\ 1 & k \in \mathbb{N}\setminus U \\ \varepsilon_U & k = \infty \end{cases}$$

(ε_U denotes the U'th unit vector in X).

$X := \overline{\text{lin}}\{hx_U \mid h \in CK,\ U \in \mathfrak{U}\}$ is a function module in $\prod_{k\in K}^{\infty} X_k$ (the proof is left to the reader).

We claim that $Z(X) = \{M_h \mid h \in CK\}$. For $T \in Z(X)$ we have $T = M_\alpha$ for a suitable bounded function $\alpha: K \to \mathbb{K}$ (this follows from prop. 4.7(vi) and $Z(X_\infty) = \mathbb{K}\,\text{Id}$; see prop. 5.1(vi)). We have to show that α is continuous. Without loss of generality we may assume that $\alpha(\infty) = 0$ (otherwise we consider $T - \alpha(\infty)\text{Id}$). α is trivially continuous at every $k \in K$ such that $k \neq \infty$. Suppose that α is not continuous at ∞ , that is there is an $\varepsilon_o > 0$ such that $\Delta := \{ n \mid n \in \mathbb{N},\ |\alpha(n)| \geq \varepsilon_o\}$ is infinite. We decompose $\mathbb{N}$ as $\mathbb{N} = \Delta_o \dot{\cup} \Delta_1 \dot{\cup} \Delta_2$, where Δ_1 and Δ_2 are infinite and $\Delta_1 \cup \Delta_2 = \Delta$. Since $\mathfrak{U}$ is an ultrafilter, we have $\Delta_o \cup \Delta_1 \in \mathfrak{U}$ or $\Delta_2 \in \mathfrak{U}$. In the first case the function $\|(x_{\Delta_o \cup \Delta_1})(\cdot)\|$ is not upper semicontinuous: it vanishes at ∞ but its value is at least ε_o at every $n \in \Delta_2$. This contradicts $TX = M_\alpha X \subset X$.

Similarly it can be shown that $\Delta_2 \in \mathfrak{U}$ also leads to a contradiction.

Thus we have proved that the identical representation of X is maximal, and it remains to show that X does not have the local cns property. We will prove that ∞ fails to have a local cns. Otherwise, there would be $h_1 x_{U_1}, \ldots, h_m x_{U_m}$ such that $\max_i \|h_i(k) x_{U_i}(k)\|$ > 0 in a neighbourhood of ∞. But this is not possible, since $U_1 \cap \ldots \cap U_m \in \mathfrak{U}$ and every $U \in \mathfrak{U}$ is infinite, i.e. there are infinitely many $k \in \mathbb{N}$ such that $x_{U_1}(k) = \ldots = x_{U_m}(k) = 0$. □

The following corollary asserts that centralizer-norming systems exist if X and Z(X) are separable:

<u>9.10 Corollary</u>: Let X be a separable Banach space. Then X has a cns iff Z(X) is separable

<u>Proof</u>: Suppose that $x_1, \ldots, x_n$ is a cns of X. Then $T \mapsto (Tx_1, \ldots, Tx_n)$ is an isomorphism from Z(X) onto a subspace of $\prod_{i=1}^{n}{}^{\infty} X_i$ ($X_i := X$ for $i=1,\ldots,n$) so that Z(X) is separable if X is separable.

Conversely, suppose that Z(X) is separable. Since $Z(X) \cong CK_X$ this implies that K_X is metrizable so that every $k_o \in K_X$ satisfies condition (ii) of prop. 9.9.

X is separable so that, by prop. 9.9, every $k_o \in K_X^*$ has a local cns. Thus it suffices to show that $K_X = K_X^*$ (recall that X has a cns iff every $k_o \in K_X$ has a local cns).

Suppose that there is a $k_o \in K_X \setminus K_X^*$. We choose a disjoint sequence $(U_n)_{n \in \mathbb{N}}$ of open subsets of K_X such that $U_n \neq \emptyset$ and $k_o \notin U_n \subset V_n$, where $V_1 \supset V_2 \supset \ldots$ is a basis of the neighbourhood system of k_o. For $n \in \mathbb{N}$, let $h_n \in CK_X$ be a function such that $\|h_n\| = 1$ and $h_n|_{K_X \setminus U_n} = 0$. Then $M_{(\sum_n \xi_n h_n)} \in Z(X)$ for every $(\xi_1, \xi_2, \ldots) \in m$ (this follows from $k_o \notin K_X^*$), and $\|M_{(\sum_n \xi_n h_n)}\| = \|(\xi_1, \xi_2, \ldots)\|$. Thus Z(X) contains a nonseparable subspace, a contradiction. □

10. M-structure of $C_o(M,X)$

Throughout this chapter X will be a fixed nonzero Banach space and M a fixed nonempty locally compact Hausdorff space. We will discuss systematically the M-structure properties of $C_o(M,X)$ (section A: M-ideals ; section B: the centralizer ; section C: function module representations).

Function module techniques are essential for our considerations (in fact, some of the results are corollaries of theorems in chapter 4). Sections B and C depend on some facts concerning centralizer-norming systems and spaces with the local cns property.

A. M-ideals and M-summands in $C_o(M,X)$

10.1 Proposition:

(i) Let $Y \subset C_o(M,X)$ be a closed subspace and $Y_v := \{ f(v) \mid f \in Y\}$ for $v \in M$. Then Y is an M-ideal iff $(C_oM)Y \subset Y$ and the Y_v are M-ideals in X

(ii) Let $(\tilde{Y}_v)_{v \in M}$ be a family of M-ideals in X such that

(∗) for each $v_o \in M$ and each $x_o \in \tilde{Y}_{v_o}$ there is an $f \in C_o(M,X)$ such that $f(v) \in \tilde{Y}_v$ for every $v \in M$ and $f(v_o) = x_o$.

Then there is an M-ideal Y of $C_o(M,X)$ such that $\{ f(v) \mid f \in Y \} = \tilde{Y}_v$ for every $v \in M$

(iii) The M-ideals Y of $C_o(M,X)$ are in one-to-one correspondence with the families $(Y_v)_{v \in M}$ of M-ideals of X which satisfy (∗)

Proof:

(i) This is a special case of prop. 4.9(ii); we only have to represent $C_o(M,X)$ as a function module with base space $K := \beta M$ and component spaces $X_k := X$ (for $k \in M$) and $X_k := \{0\}$ (for $k \in \beta M \setminus M$) as in example 2 on p. 78. Note that $(C_oM)Y \subset Y$ implies that $(C^bM)Y \subset Y$ so that Y is a CK-module.

(ii) We define $Y := \{f \mid f \in C_o(M,X),\ f(v) \in \tilde{Y}_v$ for every $v \in M\}$. Y is a closed subspace of $C_o(M,X)$ and $(C_oM)Y \subset Y$. (*) implies that $\tilde{Y}_v = \{f(v) \mid f \in Y\}$ for every $v \in M$ so that, by (i), Y is an M-ideal.

(iii) If Y is an M-ideal then the family $(Y_v)_{v \in M}$ (with Y_v as in (i)) obviously satisfies (*). Y can be reconstructed from this family since, for $f \in C_o(M,X)$, $f \in Y$ iff $f(v) \in Y_v$ for every $v \in M$ (this is a special case of prop. 4.9(i)). On the other hand, it is clear that for every family $(Y_v)_{v \in M}$ such that (*) is satisfied the Y_v are determined by the M-ideal $\{f \mid f \in C_o(M,X),\ f(v) \in Y_v$ for every $v \in M\}$. □

<u>10.2 Corollary</u>:

(i) Suppose that X has no nontrivial M-ideals. Then the M-ideals of $C_o(M,X)$ are precisely the subspaces $Y_C := \{f \mid f \in C_o(M,X), f|_C = 0\}$ ($C \subset M$, C a closed subset)

(ii) If X has no nontrivial M-summands, then the M-summands of $C_o(M,X)$ are just the subspaces Y_C, where $C \subset M$, C clopen

<u>Proof</u>:

(i) Y_C is a C_oM-module and $(Y_C)_v = \{0\}$ or $= X$ (for $v \in C$ or $v \in M \setminus C$) so that Y_C is an M-ideal by prop. 10.1(i).
Conversely, let Y be an M-ideal in $C_o(M,X)$ and $C := \{v \mid v \in M,\ Y_v = \{0\}\}$. C is obviously closed and we have $Y_C = Y$ (since $Y_v = X$ for $v \in M \setminus C$; recall that $f \in Y$ iff $f(v) \in Y_v$ for every $v \in M$).

(ii) It is clear that Y_C is an M-summand for every clopen subset C of M. Conversely, if Y is an M-summand, it follows in a similar manner to the proof of (i) that there are closed subsets $C, C^{\perp}$ of M such that $Y = Y_C$, $Y^{\perp} = Y_{C^{\perp}}$. $Y + Y^{\perp} = C_o(M,X)$ and $Y \cap Y^{\perp} = \{0\}$ imply that $C \cap C^{\perp} = \emptyset$ and $C \cup C^{\perp} = M$, i.e. C is clopen. □

<u>Note</u>: If X has only a finite number of M-ideals, then the families $(Y_v)_{v \in M}$ such that (*) (of prop. 10.1) is satisfied can easily be

described:

$(Y_v)_{v\in M}$ satisfies ($*$) iff for every $v_o \in M$ there is a neighbourhood V of v_o such that $Y_v \supset Y_{v_o}$ for every $v \in V$.

[Suppose that $(Y_v)_{v\in M}$ satisfies ($*$) and that $v_o \in M$. We choose $x_1,\ldots,x_n \in Y_{v_o}$ such that $\max_i d(x_i,Y_o) < 1$ implies $Y_{v_o} \subset Y_o$ for every M-ideal Y_o of X (such a family exists since there are only finitely many M-ideals in X). But then, if $f_1,\ldots,f_n \in C_o(M,X)$ are chosen such that $f_i(v) \in Y_v$ and $f_i(v_o) = x_i$ for $v \in M$ and $i=1,\ldots,n$, we have $Y_v \supset Y_{v_o}$ for v in the open set

$\{v \mid v \in M, \|f_i(v) - f_i(v_o)\| < 1 \text{ for } i=1,\ldots,n\}$.

The reverse implication is obvious.]

It has further been shown ([10], th. 3.8(i)) that when X only has finitely many M-ideals the M-ideals in $C_o(M,X)$ are in one-to-one correspondence with certain lattice homomorphisms from the lattice of M-ideals of X into the lattice of closed subsets of M. Similar results can be proved for M-summands ([10], th. 3.8(ii)).

B. The centralizer of $C_o(M,X)$

10.3 Proposition:

$$Z(C_o(M,X)) = \left\{ \prod_{v\in M} T_v \;\middle|\; \begin{array}{l} v \mapsto T_v \text{ is a norm bounded continuous map from} \\ M \text{ into } Z(X) \text{ with the strong operator topology} \end{array} \right\}$$

($\prod_{v\in M} T_v$ denotes the operator $[(\prod_{v\in M} T_v)f](v) := T_v(f(v))$)

Proof: Suppose that $T \in Z(C_o(M,X))$. Then $T = \prod_{v\in M} T_v$, where $T_v \in Z(X)$ and $\|T_v\| \leq \|T\|$ for every $v \in M$ (see prop. 4.7(iv)). For $x \in X$ and $v_o \in M$ we choose $f \in C_o(M,X)$ such that $f(v) = x$ for v in a neighbourhood V of v_o. Then $(Tf)(v) = T_v x$ for $v \in V$, i.e. $v \mapsto T_v x$ is continuous at v_o. Conversely, suppose that $v \mapsto T_v$ is a norm bounded map from M to $Z(X)$

and that $v \mapsto T_v x$ is continuous for every $x \in X$. Similarly to the proof of lemma 8.1 it can be shown that $(\prod_{v \in M} T_v)(C_o(M,X)) \subset C_o(M,X)$.
$v \mapsto T_v^* x$ is also continuous on M for every $x \in X$

(note that $\|Tx\| = \|T^*x\|$ for $T \in Z(X)$ and $x \in X$; this follows from the fact that $\|Tx\| = p(Tx)$, $\|T^*x\| = \tilde{p}(T^*x)$ for suitable $p, \tilde{p} \in E_X$: $\|Tx\| = |p(Tx)| = |a_T(p)p(x)| = |\overline{a_T(p)}p(x)| = |p(T^*x)| \le \|T^*x\| = |\tilde{p}(T^*x)| = |a_{T^*}(\tilde{p})\tilde{p}(x)| = |\overline{a_T(\tilde{p})}\tilde{p}(x)| = |\tilde{p}(Tx)| \le \|Tx\|$).

Thus $(\prod_{v \in M} T_v^*)(C_o(M,X)) \subset C_o(M,X)$ and consequently $\prod_{v \in M} T_v \in Z(C_o(M,X))$ (prop. 4.7(iv)). □

<u>10.4 Corollary</u>: For $h \in C_o M$ and $T \in Z(X)$ we define $h \otimes T$: $C_o(M,X) \to C_o(M,X)$ by $[(h \otimes T)f](v) := h(v)T(f(v))$ (i.e. $h \otimes T = \prod_{v \in M}(h(v)T)$). Then $Z(C_o(M,X))$ is the strong operator closure of $C_o M \otimes Z(X) := \text{lin}\{h \otimes T \mid h \in C_o M,\ T \in Z(X)\}$

<u>Proof</u>: Since $v \mapsto h(v)T$ is continuous for $h \in C_o M$ and $T \in Z(X)$ we have $C_o M \otimes Z(X) \subset Z(C_o(M,X))$. Thus, by prop. 4.21, the strong operator closure of $C_o M \otimes Z(X)$ is contained in $Z(C_o(M,X))$.
Now suppose that $T \in Z(C_o(M,X))$, $f_1,\dots,f_n \in C_o(M,X)$ and $\varepsilon > 0$ are given. Choose a compact set $C \subset M$ such that $\|f_i|_{M \setminus C}\| \le \varepsilon$ for $i=1,\dots,n$. By prop. 10.3 there is a family $(T_v)_{v \in M}$ in $Z(X)$ such that $T = \prod_{v \in M} T_v$ and $\sup_{v \in M}\|T_v\| = \|T\|$. For $v \in C$ we have $(Tf_i)(v) = T_v(f_i(v))$ so that, since the f_i and the Tf_i are continuous, there is an open relatively compact neighbourhood U_v of v such that $\|(Tf_i)(v') - T_v(f_i(v'))\| \le \varepsilon$ (all $v' \in U_v$, all $i \in \{1,\dots,n\}$). Choose $v_1,\dots,v_m \in C$ such that $C \subset U_{v_1} \cup \dots \cup U_{v_m}$ and a partition of unity subordinate to this cover of C as in prop. 0.3. It is clear from the construction that

$$\|(Tf_i - \sum_{j=1}^{m}(h_j \otimes T_{v_j})f_i)(v)\| \le \begin{cases} 2\varepsilon\|T\| & \text{if } v \notin C \\ \varepsilon & \text{if } v \in C \end{cases}$$

for $i=1,\dots,n$, i.e. $\|Tf_i - \sum_{j=1}^{m}(h_j \otimes T_{v_j})f_i\| \le \varepsilon \max\{1, 2\|T\|\}$.
This proves that T lies in the strong operator closure of $C_o M \otimes Z(X)$. □

Remark: Cor. 10.4 is a special case of the following result: for arbitrary Banach spaces X, Y $Z(X \hat{\otimes}_{\varepsilon} Y)$ is the strong operator closure of $Z(X) \otimes Z(Y)$ ([13],[88]; in both papers only real spaces are considered, the complex case follows from the fact that $Z(X) = Z(X_{\mathbb{R}}) + iZ(X_{\mathbb{R}})$).

By prop. 10.3 the centralizer of $C_o(M,X)$ can be identified with the space of bounded continuous maps from M into Z(X) with the strong operator topology. This is in general not sufficient to construct $K_{C_o(M,X)}$ from M and K_X. We will treat this problem in more detail in section C.

The following corollary asserts that M and K_X determine $K_{C_o(M,X)}$ provided that X has a cns.

10.5 Corollary: Suppose that X has a cns (see def. 9.1). Then

(i) $Z(C_o(M,X)) \cong C^b(M,Z(X))$ (Z(X) provided with the norm topology)

(ii) $K_{C_o(M,X)} \cong \beta(M \times K_X)$

Proof:

(i) This follows at once from prop. 10.3 and the fact that $\| \prod_{v \in M} T_v \| = \sup \|T_v\|$

(ii) Since M is locally compact we have $C^b(M,CK_X) \cong C^b(M \times K_X)$. Accordingly $C(K_{C_o(M,X)}) \cong Z(C_o(M,X)) \cong C^b(M,Z(X)) \cong C^b(M,CK_X) \cong C^b(M \times K_X) \cong C(\beta(M \times K_X))$ so that, by th. 7.1, $K_{C_o(M,X)} \cong \beta(M \times K_X)$. □

C. Function module representations of $C_o(M,X)$

The aim of this section is to construct (maximal) function module representations of $C_o(M,X)$ from (maximal) function module representations of X. This is an essential prerequisite if we wish to apply M-structure techniques to generalizations of th. 7.1 (cf. the introduction to part II, p. 136).

Suppose that X has been represented as a function module. As a first step we consider a natural candidate for a representation of $C_o(M,X)$ associated with the representation of X. Secondly we show that this candidate is in fact a maximal function module representation of $C_o(M,X)$ provided that the representation of X is maximal and X has the local cns property.

<u>10.6 Definition</u>: Let X be represented as a function module. For simplicity we will asume that X is identified with $\rho(X)$ (ρ as in def. 4.12), i.e. that X <u>is</u> a function module with base space K and component spaces $(X_k)_{k\in K}$.

Our candidate for a suitable function module representation of $C_o(M,X)$ is constructed as follows:

base space: $K_M := \beta(M\times K^*)$, where $K^* := \{k \mid k\in K,\ X_k \neq \{0\}\}$

component spaces: $X^M_{\tilde{k}} := \begin{cases} X_k & \text{if } \tilde{k} = (v,k)\in M\times K^* \\ \{0\} & \text{otherwise} \end{cases}$ (all $\tilde{k}\in K_M$)

function module: $X_M := \{\tilde{f} \mid f\in C_o(M,X)\} \subset \prod^{\infty}_{\tilde{k}\in K_M} X^M_{\tilde{k}}$

where $\tilde{f}(\tilde{k}) := \begin{cases} (f(v))(k) & \text{if } \tilde{k} = (v,k)\in M\times K^* \\ 0 & \text{otherwise} \end{cases}$.

We will prove at once (th. 10.8) that $(K_M, (X^M_{\tilde{k}})_{\tilde{k}\in K_M}, X_M)$ is a function module and that $X_M \cong C_o(M,X)$, i.e. every function module representation of X gives rise to a function module representation of $C_o(M,X)$. $(K_M, (X^M_{\tilde{k}})_{\tilde{k}\in K_M}, X_M)$ will be called <u>the representation of $C_o(M,X)$ associated with</u> $(K,(X_k)_{k\in K},X)$.

<u>Remark</u>: The construction may be visualized as follows: $(X^M_{\tilde{k}})_{\tilde{k}\in K_M}$ is essentially the union of card M copies of the family $(X_k)_{k\in K}$, and the essential part of $\tilde{f}$ is obtained from f by regarding every $f(v)\in X$ as an element of $\prod^{\infty}_{k\in K^*} X_k$:

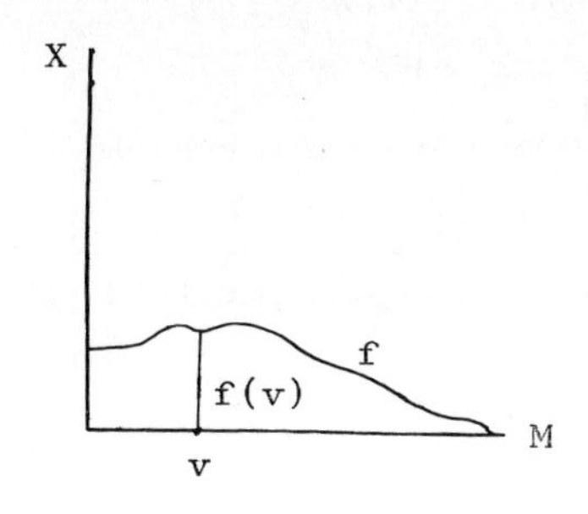

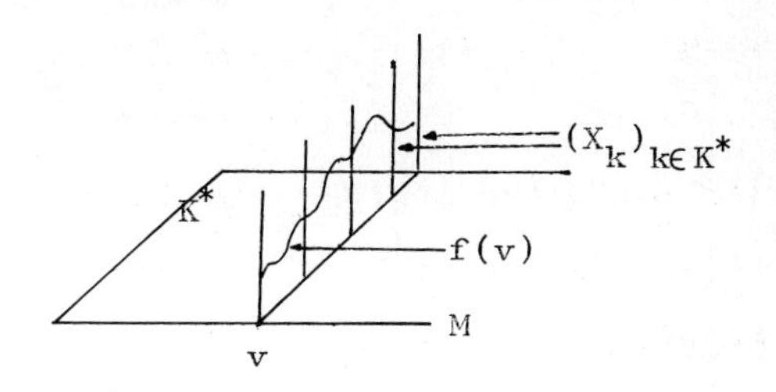

fig. 23

We note that the natural isomorphism from $C(K,CL)$ onto $C(K\times L)$ (K and L compact Hausdorff spaces) is similarly defined.

10.7 Lemma: For $h_o \in CK_M$, compact sets $A \subset M$ and $B \subset K^*$, and $\varepsilon > 0$ there are $h_1,\dots,h_n \in C_oM$, $g_1,\dots,g_n \in CK$ such that

$$\left| \sum_{i=1}^{n} h_i(v)g_i(k) - h_o(v,k) \right| \le \begin{cases} \varepsilon & \text{if } (v,k) \in A\times B \\ 2(\|h_o\|+\varepsilon) & \text{if } (v,k) \in (M\times K^*) \setminus (A\times B) \end{cases}$$

Proof:

1. Choose $\tilde{h}_1,\dots,\tilde{h}_n \in CA$, $\tilde{g}_1,\dots,\tilde{g}_n \in CB$ such that $\| \sum_{i=1}^{n} \tilde{h}_i \otimes \tilde{g}_i - h_o|_{A\times B}\| \le \varepsilon$ (where $(h\otimes g)(v,k) := h(v)g(k)$). This is possible since $\text{lin}\{h\otimes g \mid h \in CA,\ g \in CB\}$ ($= CA \otimes CB$) is dense in $C(A\times B)$.
2. Extend $\tilde{h}_i$ (and $\tilde{g}_i$) as a bounded continuous function to all of M (to K); the extensions will be denoted by h_i^* (by g_i^*).
3. We have $\| \sum_{i=1}^{n} h_i^* \otimes g_i^* |_{A\times B}\| = \| \sum_{i=1}^{n} \tilde{h}_i \otimes \tilde{g}_i \| \le \|h_o\|+\varepsilon$ so that there are compact neighbourhoods $\hat{A}$ of A and $\hat{B}$ of B such that $\| \sum_{i=1}^{n} h_i^* \otimes g_i^* |_{\hat{A}\times\hat{B}}\| \le \|h_o\|+2\varepsilon$.
4. Choose continuous functions $h^*: M \to [0,1]$, $g^*: K \to [0,1]$ for which $h^*|_{M\setminus\hat{A}} = 0$, $h^*|_A = 1$, $g^*|_{K\setminus\hat{B}} = 0$, $g^*|_B = 1$ and define $h_i^* := h_i^*\, h^*$, $g_i := g_i^*\, g^*$ for $i=1,\dots,n$.

It is obvious that these functions have the claimed properties. □

<u>10.8 Theorem</u>: (i) $(K_M, (X^M_{\tilde{k}})_{\tilde{k}\in K_M}, X_M)$ is a function module

(ii) $\rho_M : C_o(M,X) \to X_M$ is an isometrical isomorphism
$f \mapsto \tilde{f}$

(i.e. $[\rho_M, (K_M, (X^M_{\tilde{k}})_{\tilde{k}\in K_M}, X_M)]$ is a function module representation of $C_o(M,X)$)

<u>Proof</u>:

$$
\begin{aligned}
\text{(ii)}\ \|\tilde{f}\| &= \sup\{\|\tilde{f}(v,k)\| \mid (v,k) \in M \times K^*\} \\
&= \sup_{v\in M} \sup_{k\in K^*} \|f(v,k)\| \\
&= \sup_{v\in M} \|f(v)\| \\
&= \|f\| \qquad \text{for every } f \in C_o(M,X).
\end{aligned}
$$

(i) We have to show that 4.1(i) - (iv) are satisfied. X_M is obviously a closed subspace of $\prod^{\infty}_{\tilde{k}\in K_M} X^M_{\tilde{k}}$ since $\|\tilde{f}\| = \|f\|$ for every $f \in C_o(M,X)$ and $C_o(M,X)$ is complete.

<u>4.1(ii)</u>: Suppose that $f \in C_o(M,X)$ and that $\tilde{k}_o \in K_M$.
At first we will assume that $\tilde{k}_o = (v_o,k_o) \in M \times K^*$. For $\varepsilon > 0$ we choose neighbourhoods V ($\subset M$) of v_o and U ($\subset K^*$) of k_o such that $\|f(v) - f(v_o)\| \leq \varepsilon$ for $v \in V$ and $\|(f(v_o))(k)\| \leq \|(f(v_o))(k_o)\| + \varepsilon$ for every $k \in U$. It follows that $\|\tilde{f}(\tilde{k})\| \leq \|\tilde{f}(\tilde{k}_o)\| + 2\varepsilon$ for every $\tilde{k} \in W$, where W is any neighbourhood of $\tilde{k}_o$ (neighbourhood in K_M) such that $W \cap (M \times K^*) = V \times U$.
Now suppose that $\tilde{k}_o \in K_M \setminus (M \times K^*)$ and that $\varepsilon > 0$. Since f is continuous and vanishes at infinity and since the functions $k \mapsto \|(f(v))(k)\|$ are upper semicontinuous for every $v \in M$ we obtain, by a simple compactness argument, compact subsets $A \subset M$ and $B \subset K^*$ such that $\|\tilde{f}(v,k)\| \leq \varepsilon$ whenever $(v,k) \notin A \times B$.
It follows that $W := K_M \setminus (A \times B)$ is a neighbourhood of $\tilde{k}_o$ such that $\|\tilde{f}(\tilde{k})\| \leq \varepsilon$ for every $\tilde{k} \in W$ (note that $A \times B$, as a compact subset of $M \times K^*$, is also compact in K_M).
Thus $\tilde{k} \mapsto \|\tilde{f}(\tilde{k})\|$ is upper semicontinuous.
<u>4.1(i)</u>: For $f \in C_o(M,X)$, $h_o \in CK_M$, and $\varepsilon > 0$ we choose compact subsets

$A \subset M$ and $B \subset K^*$ such that $\|\tilde{f}(\tilde{k})\| \leq \varepsilon$ for $\tilde{k} \notin A \times B$ and functions $h_1,\ldots,h_n \in C_oM$, $g_1,\ldots,g_n \in CK$ for h_o, A, B, ε as in the preceding lemma ($\{\tilde{k} \mid \|\tilde{f}(\tilde{k})\| \geq \varepsilon\}$ is compact by the first part of the proof). The functions $(h_i \otimes M_{g_i})f$ are contained in $C_o(M,X)$ by cor. 10.4, and it follows from the construction that $\|h_o\tilde{f} - (\sum_{i=1}^{m}(h_i \otimes M_{g_i})f)^{\sim}\| \leq \max\{\varepsilon\,\|f\|, 2\,\varepsilon(\|h_o\| + \varepsilon)\}$. Accordingly we have $h_o\tilde{f} \in (X_M)^{-} = X_M$.

4.1(iii): For $\tilde{k} = (v,k) \in M \times K^*$ and $x_k \in X_k = X^M_{(v,k)}$ choose $x \in X$ such that $x(k) = x_k$ and $f_o \in C_o(M,X)$ such that $f_o(v) = x$. It follows that $\tilde{f}_o(\tilde{k}) = x_k$ which proves that $X^M_{\tilde{k}} = \{\tilde{f}(\tilde{k}) \mid \tilde{f} \in X_M\}$. This is also true (by definition) for $\tilde{k} \in K_M \setminus (M \times K^*)$.

4.1(iv): We have $\{\tilde{k} \mid X^M_{\tilde{k}} \neq \{0\}\} = M \times K^*$, and this set is dense in K_M. □

10.9 Theorem: (X as in def. 10.6). Suppose that every operator in $Z(X)$ is the multiplication operator M_α associated with a bounded scalar-valued function α on K. Then X_M ($\cong C_o(M,X)$) has the same property: For every operator $T: X_M \to X_M$ we have $T \in Z(X_M)$ iff there is a bounded function $\alpha: K_M \to \mathbb{K}$ such that $T = M_\alpha$

Proof: Suppose that $T \in Z(X_M)$. By prop. 10.3 there is a family $(T_v)_{v \in M}$ in $Z(X)$ such that $\rho_M^{-1} T \rho_M = \prod_{v \in M} T_v$ (ρ_M as in def. 10.8). By hypothesis there are functions $\alpha_v: K \to \mathbb{K}$ such that $\|\alpha_v\| = \|T_v\| \leq \|T\|$ and $T_v = M_{\alpha_v}$ (all $v \in M$); we define $\alpha: K_M \to \mathbb{K}$ by

$$\alpha(\tilde{k}) := \begin{cases} \alpha_v(k) & \text{if } \tilde{k} = (v,k) \in M \times K^* \\ 0 & \text{otherwise} \end{cases}$$

and claim that $T = M_\alpha$.

We have to show that $(T\tilde{f})(\tilde{k}) = (\alpha\tilde{f})(\tilde{k})$ for every $f \in C_o(M,X)$ and every $\tilde{k} \in K_M$. For $\tilde{k} \in K_M \setminus (M \times K^*)$ this is trivially satisfied, and for $\tilde{k} = (v,k) \in M \times K^*$ we have

$$\begin{aligned} (T\tilde{f})(\tilde{k}) &= [\rho_M(\prod_{v \in M} T_v)\, \rho_M^{-1}\tilde{f}](\tilde{k}) \\ &= [\rho_M(\prod_{v \in M} T_v) f](v,k) \\ &= [(\prod_{v \in M} T_v f)(v)](k) \\ &= [T_v f(v)](k) \\ &= \alpha_v(k)[f(v)](k) \end{aligned}$$

$= \alpha(\tilde{k})\tilde{f}(\tilde{k})$.

Conversely it is true in general that operators of the form M_α are contained in the centralizer of a function module (prop. 4.11(iii)).

□

Now suppose that the function module representation of X under consideration is maximal (since X is regarded as a function module this just means that $Z(X) = \{M_h \mid h \in CK\}$; note that $K = K_X$ in this case). It follows from the preceding theorem that the associated function module representation of $C_o(M,X)$ from def. 10.6 has the property that the centralizer consists solely of operators M_α ($\alpha: K_M \to \mathbb{K}$ a bounded function). However, in order to apply the results of chapter 4 (in particular cor. 4.17) we need a maximal function module representation of $C_o(M,X)$, i.e. we must show that α can be chosen to be continuous. Since $M_\alpha = M_h$ implies that $\alpha(\tilde{k}) = h(\tilde{k})$ for every $\tilde{k}$ such that $X^M_{\tilde{k}} \neq \{0\}$ (i.e. $\alpha|_{M\times K^*} = h|_{M\times K^*}$) we must prove that α is continuous on $M\times K^*$ whenever $M_\alpha(X_M) \subset X_M$. Our next theorem shows that this is true if X has the local cns property (def. 9.8).

<u>10.10 Theorem</u>: Suppose that X is represented as a function module and that, in addition, this representation is maximal (i.e. $Z(X) = \{M_h \mid h \in CK\}$). Then, if X has the local cns property, the representation $[\rho_M, (K_M, (X^M_{\tilde{k}})_{\tilde{k}\in K_M}, X_M)]$ of $C_o(M,X)$ (see def. 10.6) is also maximal

<u>Proof</u>: Let $T \in Z(X_M)$ be arbitrary. By th. 10.9 there exists a bounded function $\alpha: K_M \to \mathbb{K}$ such that $T_\alpha = M$.

We claim that α is continuous on $\{\tilde{k} \mid X^M_{\tilde{k}} \neq \{0\}\} = M\times K^*$. For $(v_o,k_o) \in M\times K^*$ we choose a compact neighbourhood V of v_o, a function $h_o \in C_oM$ such that $h_o|_V = 1$, a compact neighbourhood U of k_o and vectors $x_1,\dots,x_n \in X$ such that $\max_i \|x_i(k)\| \geq \delta > 0$ for every $k \in U$ (cf. def. 9.7). For i=1,...,n, the function $v \mapsto h_o(v)x_i$ ($\in C_o(M,X)$) will be denoted by $h_o \otimes x_i$, and we will write $h_o \tilde{\otimes} x_i$ instead of $(h_o \otimes x_i)^\sim$.

Now let $\varepsilon > 0$ be given. $(\alpha(v_o,k_o) - \alpha)h_o \tilde{\otimes} x_i$ is contained in X_M and vanishes at (v_o,k_o) so that, since X_M is a function module, there is a neighbourhood $\tilde{W}_\varepsilon$ of (v_o,k_o) such that $\|(\alpha(v_o,k_o) - \alpha)(h_o \tilde{\otimes} x_i)(\tilde{k})\| \leq \varepsilon$ for $\tilde{k} \in \tilde{W}_\varepsilon$ and $i=1,\dots,n$. $W_\varepsilon := \tilde{W}_\varepsilon \cap (V \times U)$ is a neighbourhood of (v_o,k_o) contained in $M \times K^*$, and for $(v,k) \in W_\varepsilon$ it follows that

$|\alpha(v_o,k_o) - \alpha(v,k)| \, |h_o(v)| \, \|x_i(k)\| \leq \varepsilon$ (all $i \in \{1,\dots,n\}$), i.e.

$|\alpha(v_o,k_o) - \alpha(v,k)| \leq \varepsilon/\delta$. Thus α is continuous at (v_o,k_o).

By the definition of $\beta(M \times K^*)$ there is an $h \in CK_M$ which extends $\alpha|_{M \times K^*}$. It is clear that $M_\alpha = M_h$. □

Remark: We note that, with $V, h_o, U, x_1, \dots, x_n$ as in the preceding proof, the $h_o \tilde{\otimes} x_1, \dots, h_o \tilde{\otimes} x_n$ are a local cns for (v_o,k_o) since $\max_i \|(h_o \tilde{\otimes} x_i)(\tilde{k})\| \geq \delta$ for $\tilde{k} \in V \times U$. Thus, since $(v_o,k_o) \in K^*_{C_o(M,X)}$ ($= M \times K^*$) is arbitrary and the representation under consideration is maximal by the preceding theorem, $C_o(M,X)$ also has the local cns property.

10.11 Corollary: If X has the local cns property, then

$K_{C_o(M,X)} \cong \beta(M \times K^*_X)$ and $K^*_{C_o(M,X)} \cong M \times K^*_X$ □

10.12 Corollary: Let X be a Banach space such that $Z(X) = \mathbb{K}\, Id$. Then the natural representation of $C_o(M,X)$ with base space $K := \beta M$ and component spaces $X_k := X$ (for $k \in M$) and $X_k := \{0\}$ (for $k \in \beta M \setminus M$) is maximal

Proof: This is clear since $K_X = K^*_X = \{1\}$. □

11. Generalizations of the Banach-Stone theorem

In this chapter we will combine results of the previous chapters to treat the problem as to whether or not a given Banach space has the (strong) Banach-Stone property.

In fact, our methods give answers to the following slightly more general question which involves two (possibly) distinct Banach spaces: Suppose that X, Y are Banach spaces and M, N locally compact Hausdorff spaces. What can be concluded from the existence of an isometric isomorphism from $C_o(M,X)$ to $C_o(N,Y)$? In particular, does it follow that $M \cong N$ (and/or that $X \cong Y$?)

(A preliminary result in this direction has been proved in th. 8.10: if $Z(X) = \mathbb{K}\,Id$, $Z(Y) = \mathbb{K}\,Id$, then $C_o(M,X) \cong C_o(N,Y)$ implies that $M \cong N$ and that $X \cong Y$.)

The work of the present chapter depends essentially on

- the fact that every Banach space admits a maximal function module representation (th. 4.14)
- the possibility of constructing maximal function module representations of $C_o(M,X)$ from those of X (where X is a Banach space with the local cns property; see section C of chapter 10)
- the fact that isometric isomorphisms between function modules can be explicitly described provided that the identical representations are maximal (cor. 4.17).

We will see in <u>section A</u> that by putting together these results carefully we get the following:

If X and Y have the local cns property and if $C_o(M,X)$ and $C_o(N,Y)$ are isometrically isomorphic, then $M \times \Delta_{P,X}$ and $N \times \Delta_{P,Y}$ are homeomorphic for every $P \in \mathfrak{P}$, where $(\Delta_{P,X})_{P \in \mathfrak{P}}$

and $(\Delta_{P,Y})_{P\in\mathfrak{P}}$ are families of certain subsets of K_X and K_Y, respectively.

In particular, X has the Banach-Stone property if there is a $\Delta_{P,X}$ consisting solely of one point.

Several examples of how to determine members of $(\Delta_{P,X})_{P\in\mathfrak{P}}$ are discussed. Roughly speaking, the better we know a maximal function module representation of X, the more elements of $(\Delta_{P,X})_{P\in\mathfrak{P}}$ we can obtain (we do not know whether it is possible in general to give an explicit description of this family).

In section B we will consider the special case of M-finite Banach spaces. We will completely classify the M-finite Banach spaces with the (strong) Banach-Stone property by means of the M-exponents, a result which contains the theorems of Jerison, Cambern, and Sundaresan mentioned in chapter 8 as special cases.

Finally, we summarize our results concerning generalizations of the Banach-Stone theorem (section C).

A. Generalizations of the Banach-Stone theorem: the case of Banach spaces with the local cns property

Let X, Y be nonzero Banach spaces and M, N nonvoid locally compact Hausdorff spaces. We will regard X and Y as function modules as in theorem 4.14 with base spaces K_X and K_Y and with component spaces $(X_k)_{k\in K_X}$ and $(Y_l)_{l\in K_Y}$, respectively.

Further, for $f\in C_o(M,X)$ and $v\in M$, $k\in K_X$ we will write $f(v,k)$ $(\in X_k)$ instead of $(f(v))(k)$. Similarly, $g(w,l)$ means $(g(w))(l)$ $(g\in C_o(N,Y)$, $w\in N$, $l\in K_Y)$.

[Note: This will be convenient in formulating the assertion of the following theorem. Our notation is essentially an identification of $f \in C_o(M,X)$ with $\tilde{f} \in X_M$ (see def. 10.6).]

11.1 Theorem: Suppose that X and Y have the local cns property and that there is an isometric isomorphism $I: C_o(M,X) \to C_o(N,Y)$. Then there are

- a homeomorphism $t: N \times K_Y^* \to M \times K_X^*$
- a family of isometric isomorphisms
  $u_{(w,l)}: X^M_{t(w,l)} \to Y^N_{(w,l)}$, all $(w,l) \in N \times K_Y^*$
  (where $X^M_{(v,k)} := X_k$ and $Y^N_{(w,l)} := Y_l$ for $(v,k) \in M \times K_X^*$ and $(w,l) \in N \times K_Y^*$)

such that $(If)(w,l) = u_{(w,l)}[(f \circ t)(w,l)]$ for $f \in C_o(M,X)$ and $(w,l) \in N \times K_Y^*$

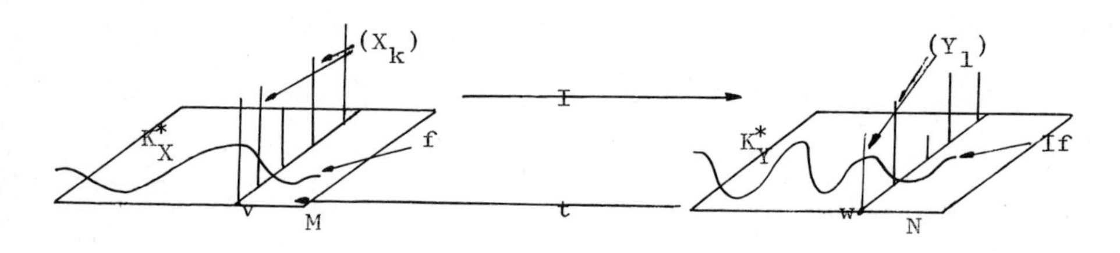

fig. 24

Proof:

We consider the function module representations $[\rho_M, (K_M, (X^M_{\tilde{k}})_{\tilde{k} \in K_M}, X_M)]$ and $[\rho_N, (K_N, (Y^N_{\tilde{l}})_{\tilde{l} \in K_N}, Y_N)]$ of $C_o(M,X)$ and $C_o(N,Y)$, respectively, as defined in def. 10.6, i.e.

$$K_M := \beta(M \times K_X^*) \qquad K_N := \beta(N \times K_Y^*)$$

$$X^M_{\tilde{k}} := \begin{cases} X_k & \text{if } \tilde{k} = (v,k) \in M \times K_X^* \\ \{0\} & \text{otherwise} \end{cases} \qquad Y^N_{\tilde{l}} := \begin{cases} Y_l & \text{if } \tilde{l} = (w,l) \in N \times K_Y^* \\ \{0\} & \text{otherwise} \end{cases}$$

$$\tilde{f}(\tilde{k}) := \begin{cases} (f(v))(k) & \text{if } \tilde{k} = (v,k) \in M \times K_X^* \\ 0 & \text{otherwise} \end{cases} \qquad \tilde{g}(\tilde{l}) := \begin{cases} (g(w))(l) & \text{if } \tilde{l} = (w,l) \in N \times K_Y^* \\ 0 & \text{otherwise} \end{cases}$$

$$\rho_M(f) := \tilde{f} \quad (\text{all } f \in C_o(M,X)) \qquad \rho_N(g) := \tilde{g} \quad (\text{all } g \in C_o(N,Y))$$

$$X_M := \text{range } \rho_M \qquad Y_N := \text{range } \rho_N \quad .$$

(We note that, with the notation from the beginning of this section, $\tilde{f}(v,k) = f(v,k)$ and $\tilde{g}(w,l) = g(w,l)$ for $(v,k) \in M \times K_X^*$ and $(w,l) \in N \times K_Y^*$.)

X and Y have the local cns property so that, by th. 10.10, these representations are maximal.

$\hat{I} := \rho_N \circ I \circ \rho_M^{-1} : X_M \to Y_N$ is an isometrical isomorphism

$$\begin{array}{ccc} X_M & \xrightarrow{\hat{I}} & Y_N \\ \uparrow \rho_M & & \uparrow \rho_N \\ C_o(M,X) & \xrightarrow{I} & C_o(N,Y) \end{array}$$

so that, by cor. 4.17, there are a homeomorphism $t : K_N \to K_M$ and a family of isometric isomorphisms $u_{\tilde{l}} : X^M_{t(\tilde{l})} \to Y^N_{\tilde{l}}$ $(\tilde{l} \in K_N)$ such that $(\hat{I}\tilde{f})(\tilde{l}) = u_{\tilde{l}}\, \tilde{f}(t(\tilde{l}))$ for every $\tilde{f} \in X_M$, $\tilde{l} \in K_N$.

Since the $X^M_{\tilde{k}}$ (the $Y^N_{\tilde{l}}$) are different from zero precisely for the $\tilde{k} \in M \times K^*_X$ (the $\tilde{l} \in N \times K^*_Y$) it follows that $t(N \times K^*_Y) = M \times K^*_X$, i.e. t induces a homeomorphism (which will also be denoted by t) from $N \times K^*_Y$ onto $M \times K^*_X$. Hence, for $f \in C_o(M,X)$ and $(w,l) \in N \times K^*_Y$

$$\begin{aligned} (If)(w,l) &= [\,(\rho_N \circ I) f\,](w,l) \\ &= [\,(\hat{I} \circ \rho_M) f\,](w,l) \\ &= u_{(w,l)}[(\rho_M(f))(t(w,l))\,] \\ &= u_{(w,l)}[f(t(w,l))] \, . \end{aligned}$$

□

<u>11.2 Corollary</u>: Let X and Y be Banach spaces with the local cns property and M and N locally compact Hausdorff spaces such that $C_o(M,X)$ and $C_o(N,Y)$ are isometrically isomorphic. Then $M \times K^*_X$ and $N \times K^*_Y$ are homeomorphic and the families $(X_k)_{k \in K_X}$ and $(Y_l)_{l \in K_Y}$ contain (up to isometric isomorphism) the same Banach spaces. □

<u>Notes</u>: 1. In particular it follows that every Banach space X such that $Z(X) = \mathbb{K}\,Id$ (so that $K_X = K^*_X = \{1\}$) has the Banach-Stone property
(in fact we have proved that such spaces even have the strong Banach-Stone property).

2. Since for spaces with the local cns property we have $K_X = K^*_X$ iff K^*_X is compact iff X has a cns the corollary

implies that:

if X and Y are Banach spaces with the local cns property, M and N nonvoid compact Hausdorff spaces such that $C(M,X) \cong C(N,Y)$, then X has a cns iff Y has a cns.

We are now going to look more carefully at the consequences of th. 11.1. We will illustrate our following definitions by a simple

example: Suppose that X is a Banach space with a cns and that in its maximal function module representation with base space K_X and component spaces $(X_k)_{k \in K_X}$ there is a distinguished point $k_o \in K_X$ such that X_{k_o} is "large" and all other X_k are "small". We assume further that the Banach space Y also has a cns and a distinguished point $l_o \in K_Y$ with "large" Y_{l_o} and "small" Y_l for $l \neq l_o$.

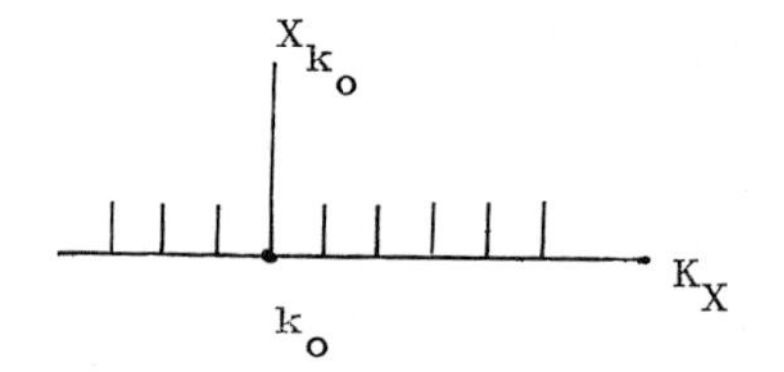

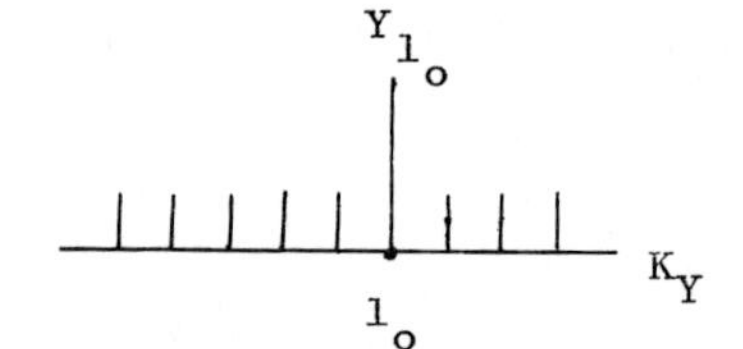

fig. 25

Then, for locally compact Hausdorff spaces M,N the families of Banach spaces in which $C_o(M,X)$ and $C_o(N,Y)$ are represented as function modules can be visualized as follows :

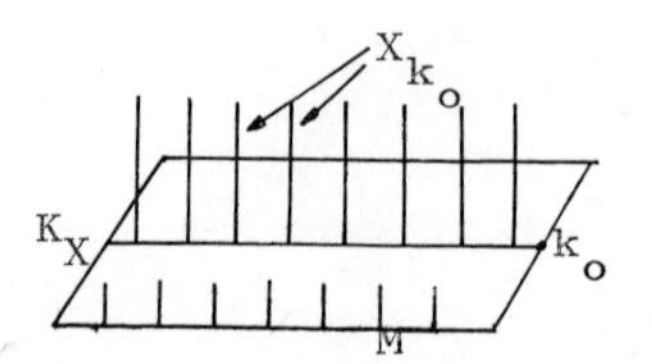

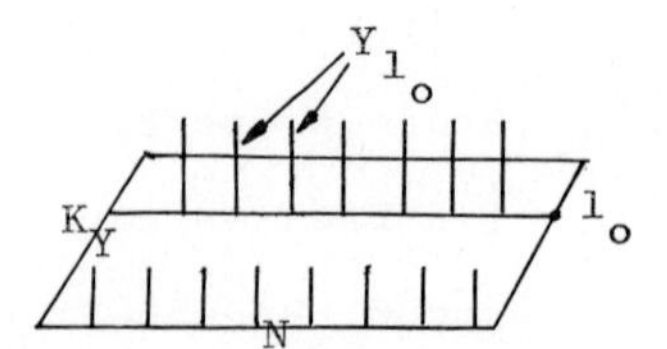

fig. 26

Now let $I:C_o(M,X) \to C_o(N,Y)$ be an isometric isomorphism and t, $(u_{(w,l)})_{(w,l)\in N\times K_Y}$ as in th. 11.1 (note that $K_Y = K_Y^*$ since Y has a cns). Since the $u_{(w,l)}:X^M_{t(w,l)} \to Y^N_{(w,l)}$ are isometric isomorphisms, $X^M_{t(w,l)}$ must be "large" iff $Y^N_{(w,l)}$ is "large", i.e.

$$\{t(w,l_o) \mid w\in N\} = \{(v,k_o) \mid v\in M\}.$$

Thus t induces a homeomorphism from $N\times\{l_o\}$ onto $M\times\{k_o\}$ so that $M \cong N$.

In particular, it follows that X has the Banach-Stone property.

In order to generalize the discussion of this example we will proceed as follows. In a <u>first step</u> we will distinguish certain subsets of K_X and K_Y (in the example: $\{k_o\}$ and $\{l_o\}$) which are, in a sense, independent of the representation of X as a function module (provided that these representations are maximal). This will be made precise in def. 11.3. <u>Secondly</u> we will discuss whether these subsets $\Delta\subset K_X$ and $\Delta'\subset K_Y$ have the property that the homeomorphism $t:N\times K_Y^* \to M\times K_X^*$ of th. 11.1 induces homeomorphisms between $N\times\Delta'$ and $M\times\Delta$ whenever M and N are locally compact Hausdorff spaces such that $C_o(M,X)$ and $C_o(N,Y)$ are isometrically isomorphic (see def. 11.5: <u>hereditary function module properties</u>).

<u>11.3 Definition</u>: A <u>function module property</u> is a rule [1] P which assigns to every function module $(K,(X_k)_{k\in K},X)$ such that $Z(X) = \{M_h \mid h\in CK\}$ a subset $P(K,(X_k)_{k\in K},X)$ of K^* $(:= \{k \mid k\in K,\ X_k \neq \{0\}\})$ such that the following holds:

If $(K,(X_k)_{k\in K},X)$ and $(L,(Y_l)_{l\in L},Y)$ are function modules such that $Z(X) = \{M_h \mid h\in CK\}$ and $Z(Y) = \{M_g \mid g\in CL\}$ and if $I:X\to Y$ is an isometrical isomorphism, then $t(P(L,(Y_l)_{l\in L},Y)) = P(K,(X_k)_{k\in K},X)$,

---

[1] We omit to give a definition using categories and functors; this would be more precise but, on the other hand, much more complicated.

where t is the homeomorphism from L to K as in cor. 4.17. Thus a function module property is a rule which assigns to every function module (for which the identical representation is maximal) a subset of the base space in such a way that only "essential" properties of function modules are used to determine this subset.

Examples:

There are a number of topological and/or functional analytical properties of the base space and/or the component spaces which give rise to function module properties:

1. Let X_o be a fixed nonzero Banach space. Then $P^1_{X_o}(K,(X_k)_{k\in K},X) := \{k \mid k\in K,\ X_k \cong X_o\}$ defines a function module property

   Proof: Let $I:X \to Y$ and $t:L \to K$ be as in cor. 4.17. Then

$$\begin{aligned} P^1_{X_o}(K,(X_k)_{k\in K},X) &= \{k \mid k\in K^*,\ X_k \cong X_o\} \\ &= \{t(l) \mid l\in L^*,\ X_{t(l)} \cong X_o\} \\ &= \{t(l) \mid l\in L^*,\ Y_l \cong X_o\} \text{ (since } X_{t(l)}\cong Y_l) \\ &= t(\{l \mid l\in L^*,\ Y_l \cong X_o\}) \\ &= t(P^1_{X_o}(L,(Y_l)_{l\in L},Y)) . \end{aligned}$$

   Notes: 1. For the distinguished point k_o and the "large" Banach space $X_{k_o} =: X_o$ of the preceding example we have $P^1_{X_o}(K,(X_k)_{k\in K},X) = \{k_o\}$.

   2. $X_o = \{0\}$ is not admissable since $P^1_{X_o}(K,(X_k)_{k\in K},X)$ must be a subset of K^*.

2. Let c be a fixed cardinal number. Then $P^2_c(K,(X_k)_{k\in K},X) :=$

   $\{k \mid k\in K^*$, the neighbourhood system of k has a basis consisting of at most c elements$\}$

   defines a function module property

   Proof: This is obvious.

3. For $\alpha\in[0,1]$ we define P^3_α by $P^3_\alpha(K,(X_k)_{k\in K},X) :=$

   $\{k_o \mid k_o\in K^*,\ \underline{\lim}_{k\to k_o}\|x(k)\| = \alpha\overline{\lim}_{k\to k_o}\|x(k)\| \text{ for every } x\in X\}$;

   P^3_α is a function module property

Proof: Let $I: X \to Y$ be as in def. 11.3. Then, for $x \in X$, we have $(Ix)(l) = S_l(x(t(l)))$ (t, $(S_l)_{l\in L}$ as in cor. 4.17). It follows that

$$\underline{\lim}_{l\to l_o} \|(Ix)(l)\| = \alpha \overline{\lim}_{l\to l_o} \|(Ix)(l)\|$$

iff

$$\underline{\lim}_{l\to l_o} \|x(t(l))\| = \alpha \overline{\lim}_{l\to l_o} \|x(t(l))\|$$

iff

$$\underline{\lim}_{k\to t(l_o)} \|x(k)\| = \alpha \overline{\lim}_{k\to t(l_o)} \|x(k)\|$$

so that $P^3_\alpha(K,(X_k)_{k\in K},X) = t(P^3_\alpha(L,(Y_l)_{l\in L},Y))$.

We give without proof some further examples of function module properties:

- $P^4_{\text{continuous}}(K,(X_k)_{k\in K},X) := \left\{ k_o \,\middle|\, \begin{array}{l} k_o \in K^*, \text{ there is an } x_o \in X \text{ such that} \\ x_o(k) \neq 0 \text{ for } k \text{ in a neighbourhood} \\ \text{of } k_o \text{ and } \|x_o(\cdot)\| \text{ is continuous at } k_o \end{array} \right\}$
- $P^5_{\text{isolated}}(K,(X_k)_{k\in K},X) := \{k \mid k \text{ is an isolated point of } K^*\}$
- $P^6_{\text{connected}}(K,(X_k)_{k\in K},X) := \{k \mid k \in K^*, k \text{ has a connected neighbourhood}\}$
- $P^7_{\text{finite-dimensional}}(K,(X_k)_{k\in K},X) :=$
  $\{k \mid k \in K^*, X_k \text{ is finite-dimensional}\}$
- $P^8_{\text{reflexive}}(K,(X_k)_{k\in K},X) := \{k \mid k \in K^*, X_k \text{ is reflexive}\}$

As an example of a definition which is not a function module property we mention

$$P(K,(X_k)_{k\in K},X) := \left\{ k_o \,\middle|\, \begin{array}{l} k_o \in K^*, \text{there is a neighbourhood } U \text{ of } k_o \text{ and a Banach} \\ \text{space } \tilde{X} \text{ such that } X_k = \tilde{X} \text{ for every } k \in U \text{ and} \\ k \mapsto \|x(k)\| \text{ is continuous on } U \text{ for every } x \in X \end{array} \right\}$$

(this is not a function module property since the definition depends on the component spaces and not only on the isometry classes of these spaces).

It is not hard to construct new function module properties from these examples:

11.4 Lemma: Let P be a function module property and $(P_i)_{i\in I}$ a family of function module properties. We define $\text{non}\, P$, P^-, P^o, $\bigwedge_{i\in I} P_i$, and $\bigvee_{i\in I} P_i$ by

$$\text{non}\, P(K,(X_k)_{k\in K},X) := K^* \setminus P(K,(X_k)_{k\in K},X)$$

$$P^-(K,(X_k)_{k\in K},X) := (P(K,(X_k)_{k\in K},X))^-$$

$$P^o(K,(X_k)_{k\in K},X) := (P(K,(X_k)_{k\in K},X))^o$$

$$(\bigwedge_{i\in I} P_i)(K,(X_k)_{k\in K},X) := \bigcap_{i\in I} P_i(K,(X_k)_{k\in K},X)$$

$$(\bigvee_{i\in I} P_i)(K,(X_k)_{k\in K},X) := \bigcup_{i\in I} P_i(K,(X_k)_{k\in K},X) .$$

Then $\operatorname{non} P$, P^-, P^o, $\bigwedge_{i\in I} P_i$, and $\bigvee_{i\in I} P_i$ are also function module properties.

Proof: Obvious. □

The next task is to single out those function module properties P which have the property that $P(K_M,(X^M_{\tilde{k}})_{\tilde{k}\in K_M},X_M)$ can be constructed from $P(K,(X_k)_{k\in K},X)$ in a simple way provided that X has the local cns property.

11.5 Definition: Let P be a function module property. We say that P is hereditary if the following holds:

If $(K,(X_k)_{k\in K},X)$ is a function module such that $Z(X) = \{M_h \mid h \in CK\}$ and X has the local cns property, then

$$P(K_M,(X^M_{\tilde{k}})_{\tilde{k}\in K_M},X_M) = M \times P(K,(X_k)_{k\in K},X)$$

for every nonvoid locally compact Hausdorff space M ($(K_M,(X^M_{\tilde{k}})_{\tilde{k}\in K_M},X_M)$ as in def. 10.6; note that $Z(X_M) = \{M_h \mid h\in CK_M\}$ by th. 10.10 so that $P(K_M,(X^M_{\tilde{k}})_{\tilde{k}\in K_M},X_M)$ is defined)

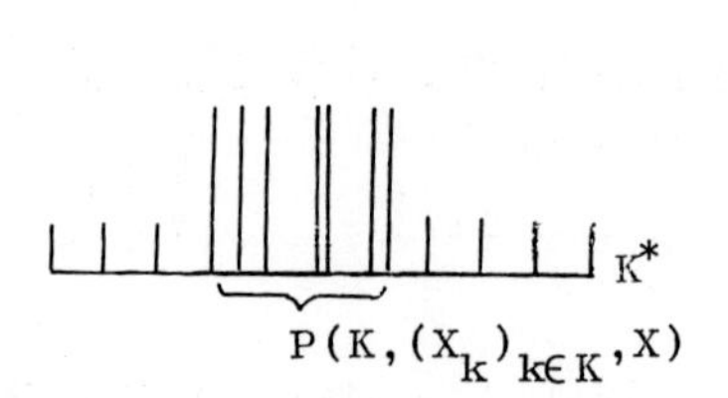

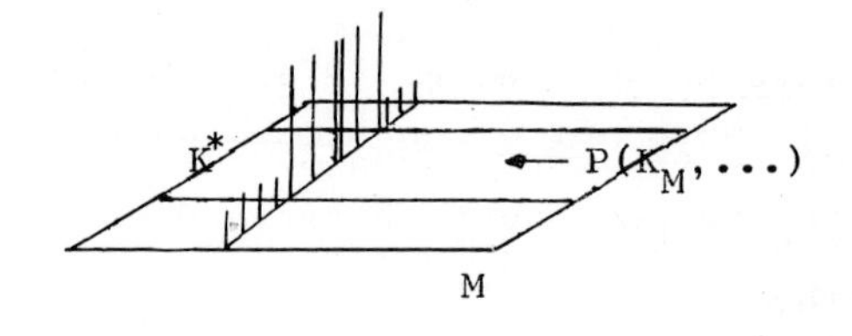

fig. 27

Thus a function module property is hereditary if it is defined by means of a topological and/or functional analytical property of $(K_M,(X^M_{\tilde{k}})_{\tilde{k}\in K_M},X_M)$ which, in a sense, only depends on the "K-coordinates" k and not on the "M-coordinates" v of the points $(v,k) \in M\times K^*$ ($= K^*_M$).

Examples:

($(K,(X_k)_{k\in K},X)$, M, $(K_M,(X^M_{\tilde{k}})_{\tilde{k}\in K_M},X_M)$ as in def. 11.5; for the definition of $P^1_{X_o}$ etc. see the examples on p. 184.)

1. $P^1_{X_o}$ is a hereditary function module property for every nonzero Banach space X_o

   Proof:
$$\begin{aligned} P^1_{X_o}(K_M,(X^M_{\tilde{k}})_{\tilde{k}\in K_M},X_M) &= \{\tilde{k} \mid \tilde{k}\in K^*_M,\ X^M_{\tilde{k}} \cong X_o\} \\ &= \{(v,k) \mid (v,k)\in M\times K^*,\ X^M_{(v,k)}(=X_k)\cong X_o\} \\ &= \{(v,k) \mid v\in M,\ k\in K^*,\ X_k \cong X_o\} \\ &= M\times P^1_{X_o}(K,(X_k)_{k\in K},X) \end{aligned}$$

2. Let c be a fixed cardinal number. We have
$$\begin{aligned} P^2_c(K_M,(X^M_{\tilde{k}})_{\tilde{k}\in K_M},X_M) &= \left\{(v,k) \;\middle|\; \begin{array}{l}(v,k)\in M\times K^* = K^*_M,\ (v,k) \text{ has a neighbourhood basis of at most } c \text{ elements}\end{array}\right\} \\ &= \left\{ v \;\middle|\; \begin{array}{l} v\in M,\ v \text{ has a neighbourhood basis of at most } c \text{ elements}\end{array}\right\} \\ &\quad \times P^2_c(K,(X_k)_{k\in K},X) \quad , \end{aligned}$$
   and this set is in general a proper subset of $M\times P^2_c(K,(X_k)_{k\in K},X)$. Thus P^2_c is not a hereditary function module property.

We note without proof that P^3_α $(\alpha\in[0,1])$, $P^4_{\text{continuous}}$, $P^7_{\text{finite-dimensional}}$, and $P^8_{\text{reflexive}}$ are hereditary function module properties whereas P^5_{isolated} and $P^6_{\text{connected}}$ are not hereditary.

11.6 Lemma: Let P, $(P_i)_{i\in I}$ be as in lemma 11.4. If P and the $(P_i)_{i\in I}$ are hereditary, then so are non$P$, P^-, P^o, $\bigwedge_{i\in I} P_i$, and $\bigvee_{i\in I} P_i$

Proof: Obvious. □

11.7 Corollary: Let $(K,(X_k)_{k\in K},X)$ be a function module such that $Z(X) = \{M_h \mid h\in CK\}$. We define $\mathfrak{B}(K,(X_k)_{k\in K},X)$ (and $\mathfrak{B}_{her}(K,(X_k)_{k\in K},X)$) to be the collection of all $P(K,(X_k)_{k\in K},X)$, where P is a function module property (a hereditary function module property). Then $\mathfrak{B}(K,(X_k)_{k\in K},X)$ and $\mathfrak{B}_{her}(K,(X_k)_{k\in K},X)$ are complete Boolean algebras of subsets of K which contain the closure and the interior of every element. □

Note: Since (hereditary) function module properties are defined by means of a class (and not of a set) of objects, it is not clear whether we are justified in regarding $\mathfrak{P}(K,(X_k)_{k\in K},X)$ and $\mathfrak{P}_{her}(K,(X_k)_{k\in K},X)$ as sets. However, this is no essential problem for the following considerations since there will be no need to treat all (hereditary) function module properties at the same time. Readers who want to avoid the use of $\mathfrak{P}(K,(X_k)_{k\in K},X)$ and $\mathfrak{P}_{her}(K,(X_k)_{k\in K},X)$ may rephrase cor. 11.7 by saying that the subsets $P(K,(X_k)_{k\in K},X)$ of K , P a (hereditary) function module property, have the property that complements, closures, etc. of such sets also have this form.

The following theorem is our most far-reaching result towards answering the question stated at the beginning of this chapter (see p. 178):

11.8 Theorem: Let X and Y be nonzero Banach spaces with the local cns property and M and N nonvoid locally compact Hausdorff spaces such that $C_o(M,X)$ and $C_o(N,Y)$ are isometrically isomorphic. Then $M\times P(K_X,(X_k)_{k\in K_X},X)$ and $N\times P(K_Y,(Y_l)_{l\in K_Y},Y)$ are homeomorphic for every hereditary function module property P (X and Y are identified with function modules $(K_X,(X_k)_{k\in K_X},X)$ and $(K_Y,(Y_l)_{l\ K_Y},Y)$ as in th. 4.14).

In particular, if there is a hereditary function module property P such that both $P(K_X,(X_k)_{k\in K_X},X)$ and $P(K_Y,(Y_l)_{l\in K_Y},Y)$ contain exactly one element, then the existence of an isometric isomorphism from $C_o(M,X)$ onto $C_o(N,Y)$ implies that M and N are homeomorphic

Proof: Let $I:C_o(M,X)\to C_o(N,Y)$ be an isometric isomorphism with t as in th. 11.1 and P a hereditary function module property. Then, by th. 11.1, def. 11.3, and def. 11.5 we have

$$\begin{aligned} t(N\times P(K_Y,(Y_l)_{l\in K_Y},Y) &= t(P(K_N,(Y^N_{\tilde{l}})_{\tilde{l}\in K_N},Y_N)) \\ &= P(K_M,(X^M_{\tilde{k}})_{\tilde{k}\in K_M},X_M) \end{aligned}$$

$$= M \times P(K_X,(X_k)_{k\in K_X},X) \; .$$

Hence $N \times P(K_Y,(Y_l)_{l\in K_Y},Y) \cong M \times P(K_X,(X_k)_{k\in K_X},X)$. □

Notes: 1. In order to apply this theorem we need to know a lot of hereditary function module properties. This motivates the search of as many such properties as possible.

2. The theorem implies that an isometric isomorphism $I:C_o(M,X) \to C_o(N,Y)$ can only exist if the sets $P(K_X,(X_k)_{k\in K_X},X)$ and $P(K_Y,(Y_l)_{l\in K_Y},Y)$ are both nonempty or both empty for every hereditary function module property P. For example, if all $k \mapsto \|x(k)\|$ are continuous (i.e. non $P^3_1(K_X,(X_k)_{k\in K_X},X) = \emptyset$) then all $l \mapsto \|y(l)\|$ must be continuous and vice versa.

In view of the generalizations of the Banach-Stone theorem the following consequence of th. 11.8 is of interest:

11.9 Theorem: Let X be a nonzero Banach space with the local cns property. Suppose that X is identified with a function module in $(K_X,(X_k)_{k\in K_X},X)$ as in th. 4.14 and that there exists a hereditary function module property P such that $P(K_X,(X_k)_{k\in K_X},X)$ contains exactly one element. Then X has the Banach-Stone property. □

It is not difficult to construct various examples of Banach spaces X for which th. 11.9 can be applied and for which the Banach-Stone property cannot be derived from the theorems of chapter 8.
As a simple example we consider the space X_s $(0<|s|\leq 1)$ which has been defined on p. 100. X_s has a cns (see p. 155) and thus the local cns property. A maximal function module representation $(K,(X_k)_{k\in K},X_s)$ has been obtained on p. 100. It is obvious that, with $\alpha = |s|$,

$$P^3_\alpha(K,(X_k)_{k\in K},X_s) = \begin{cases} K & \text{if } |s| = 1 \\ \{1\} & \text{if } |s| < 1 \end{cases}$$

so that X_s has the Banach-Stone property for every $s \in \mathbb{K}$ such that $0 < |s| < 1$. We do not know whether this is also true for $|s| = 1$.

B. Generalizations of the Banach-Stone theorem: the case of M-finite Banach spaces

We recall that a Banach space X is M-finite iff there are nonzero Banach spaces $\tilde{X}_1,\ldots,\tilde{X}_r$ and numbers $n_1,\ldots,n_r \in \mathbb{N}$ such that $Z(\tilde{X}_i) = \mathbb{K}\,\mathrm{Id}$ and $\tilde{X}_i \not\cong \tilde{X}_j$ $(i,j \in \{1,\ldots,r\},\ i \neq j)$ and $X \cong \prod_{i=1}^{r}{}^{\infty} X_i^{n_i}$. The $n_1,\ldots,n_r$ (which are, up to rearrangement, uniquely determined) are called the M-exponents of X (cf. chapter 5, section B). M-finite Banach spaces have a cns. Hence they have the local cns property, and the results of the preceding section apply.
However, in order to obtain stronger results (for example, assertions concerning the strong Banach-Stone property) it will not be sufficient to use hereditary function module properties.

At first we will discuss the structure of isometric isomorphisms between spaces $C_o(M,X)$ where X is M-finite.

11.10 Definition:

(i) Let $X_1,\ldots,X_n$ be Banach spaces and $\omega:\{1,\ldots,n\} \to \{1,\ldots,n\}$ a permutation.
By $I_\omega : \prod_{i=1}^{n}{}^{\infty} X_i \to \prod_{i=1}^{n}{}^{\infty} X_{\omega(i)}$ we denote the isometric isomorphism $(x_1,\ldots,x_n) \mapsto (x_{\omega(1)},\ldots,x_{\omega(n)})$

(ii) For Banach spaces $Y, Y_1,\ldots,Y_n$ and locally compact Hausdorff spaces M there are natural isometric isomorphisms
$C_o(M,Y^n) \cong C_o(nM,Y)$, $C_o(M, \prod_{i=1}^{n}{}^{\infty} Y_i) \cong \prod_{i=1}^{n}{}^{\infty} C_o(M,Y_i)$
(nM = the disjoint union of n copies of M).
Thus, for every M-finite Banach space $X \cong \prod_{i=1}^{r}{}^{\infty} \tilde{X}_i^{n_i}$ there is a

natural isometric isomorphism (which will be denoted by $I_{M,X}$) from $C_o(M,X)$ onto $\prod_{i=1}^{r}{}^{\infty} C_o(n_i M, X_i)$.

<u>11.11 Lemma</u>: Let $X_1,\ldots,X_r,Y_1,\ldots,Y_{\tilde r}$ be nonzero Banach spaces such that $Z(X_i) = \mathbb{K}\,\mathrm{Id}$, $Z(Y_j) = \mathbb{K}\,\mathrm{Id}$ for $i=1,\ldots,r$, $j=1,\ldots,\tilde r$ and $X_i \not\cong X_{i'}$ if $i \neq i'$ and $Y_j \not\cong Y_{j'}$ if $j \neq j'$. Further suppose that $M_1,\ldots,M_r$ and $N_1,\ldots,N_{\tilde r}$ are nonvoid locally compact Hausdorff spaces and that

$$\hat I : \prod_{i=1}^{r}{}^{\infty} C_o(M_i,X_i) \to \prod_{j=1}^{\tilde r}{}^{\infty} C_o(N_j,Y_j)$$

is an isometric isomorphism. Then $r = \tilde r$ and there are

- a permutation $\omega:\{1,\ldots,r\} \to \{1,\ldots,r\}$
- homeomorphisms $t_i: N_{\omega(i)} \to M_i$ $(i=1,\ldots,r)$
- continuous maps $u_i: N_{\omega(i)} \to [X_i, Y_{\omega(i)}]_{iso}$ $(i=1,\ldots,r)$

such that $\hat I = I_{\omega^{-1}} \circ (\prod_{i=1}^{r} I_{t_i,u_i})$ (I_{t_i,u_i} as in lemma 8.1):

$$\begin{array}{ccc}
\prod_{i=1}^{r}{}^{\infty} C_o(M_i,X_i) & \xrightarrow{\hat I} & \prod_{j=1}^{\tilde r}{}^{\infty} C_o(N_j,Y_j) \\
{\scriptstyle \prod_{i=1}^{r} I_{t_i,u_i}} \searrow & & \nearrow {\scriptstyle I_{\omega^{-1}}} \\
& \prod_{i=1}^{r}{}^{\infty} C_o(N_{\omega(i)},Y_{\omega(i)}) &
\end{array}$$

<u>Proof</u>:

At first we will prove that for every $i_o \in \{1,\ldots,r\}$ there is a $j_o \in \{1,\ldots,\tilde r\}$ such that $X_{i_o} \cong Y_{j_o}$.

For simplicity we will regard the $C_o(M_i,X_i) =: J_i$ (the $C_o(N_j,Y_j) =: J_j^*$) as subspaces of $\prod_{i=1}^{r}{}^{\infty} C_o(M_i,X_i)$ (of $\prod_{j=1}^{\tilde r}{}^{\infty} C_o(N_j,Y_j)$).

Let $i_o \in \{1,.,,,n\}$ be arbitrary. Since images of M-summands under isometric isomorphisms are also M-summands, $\hat I(J_{i_o})$ must be an M-summand in $\prod_{j=1}^{\tilde r}{}^{\infty} J_j^*$ and thus of the form $\prod_{j=1}^{\tilde r} (J_j^* \cap \hat I(J_{i_o}))$ (see the example on p. 17). Since $\hat I(J_{i_o})$ is nonzero, there must be a $j_o \in \{1,\ldots,\tilde r\}$ such that $J^* := J_{j_o}^* \cap \hat I(J_{i_o})$ is a nonzero M-summand in $J_{j_o}^*$. By prop. 5.2 and cor. 10.2(ii) there is a clopen subset C^* of N_{j_o} such that

$$J^* = \{ f \mid f \in C_o(N_{j_o},Y_{j_o}),\ f|_{C^*} = 0\} \cong C_o(N_{j_o} \setminus C^*, Y_{j_o}).$$

Similarly we obtain a clopen subset C of M_{i_o} such that $\hat{I}^{-1}(J^*) \cong C_o(M_{i_o} \setminus C, X_{i_o})$. Hence $C_o(M_{i_o} \setminus C, X_{i_o}) \cong C_o(N_{j_o} \setminus C^*, Y_{j_o})$ and therefore (by th. 8.10 and since $M_{i_o} \setminus C \neq \emptyset \neq N_{j_o} \setminus C^*$) $X_{i_o} \cong Y_{j_o}$.

Since the X_i and the Y_j are pairwise not isometrically isomorphic, the map $\omega: \{1,\ldots,r\} \to \{1,\ldots,\tilde{r}\}$, $\omega(i_o) := j_o$, is well-defined and bijective (so that, in particular, $r = \tilde{r}$). We further have $\hat{I}(J_i) = J^*_{\omega(i)}$ (since $\hat{I}(J_i) \cap J^*_j = \{0\}$ for $j \neq \omega(i)$), i.e. there are homeomorphisms $t_i: N_{\omega(i)} \to M_i$ and continuous maps $u_i: N_{\omega(i)} \to [X_i, Y_{\omega(i)}]_{iso}$ such that $\hat{I}|_{J_i} = I_{t_i,u_i}$ (th. 8.10).

It is clear from the construction that $\hat{I} = I_{\omega^{-1}} \circ (\prod_{i=1}^{r} I_{t_i,u_i})$. □

<u>11.12 Theorem</u>: Let X and Y be M-finite Banach spaces with canonical M-decompositions $X \cong \prod_{i=1}^{r\,\infty} \tilde{X}_i^{n_i}$ and $Y = \prod_{j=1}^{\tilde{r}} \tilde{Y}_j^{m_j}$, respectively, and M and N nonzero locally compact Hausdorff spaces.

(i) If $I: C_o(M,X) \to C_o(N,Y)$ is an isometrical isomorphism, then $r = \tilde{r}$ and there are

- a permutation $\omega: \{1,\ldots,r\} \to \{1,\ldots,r\}$
- homeomorphisms $t_i: m_{\omega(i)}N \to n_iM$ $(i=1,\ldots,r)$
- continuous maps $u_i: m_{\omega(i)}N \to [X_i, Y_{\omega(i)}]_{iso}$ $(i=1,\ldots,r)$

such that $I = I_{N,Y}^{-1} \circ I_{\omega^{-1}} (\prod_{i=1}^{r} I_{t_i,u_i}) \circ I_{M,X}$ ($I_{t,u}$ as in 8.1)

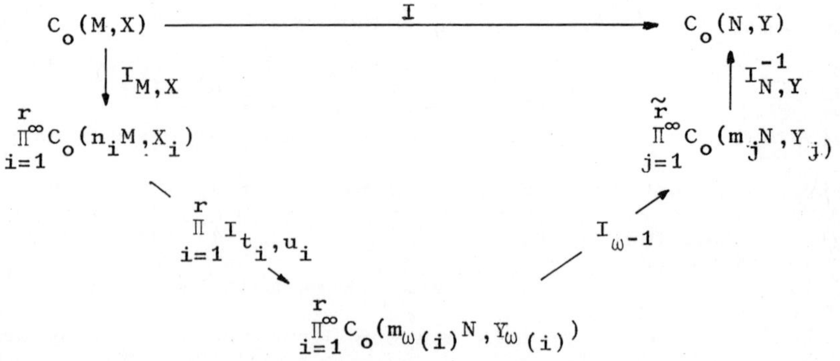

$$\begin{array}{ccc}
C_o(M,X) & \xrightarrow{\quad I \quad} & C_o(N,Y) \\
\downarrow I_{M,X} & & \uparrow I_{N,Y}^{-1} \\
\prod_{i=1}^{r\,\infty} C_o(n_iM, X_i) & & \prod_{j=1}^{\tilde{r}\,\infty} C_o(m_jN, Y_j) \\
\searrow \prod_{i=1}^{r} I_{t_i,u_i} & & \nearrow I_{\omega^{-1}} \\
& \prod_{i=1}^{r\,\infty} C_o(m_{\omega(i)}N, Y_{\omega(i)}) &
\end{array}$$

(ii) Conversely, if $r = \tilde{r}$ and ω, the t_i and the u_i are as in (i), then $I^{-1}_{N,Y} \circ I_{\omega^{-1}} \circ (\prod_{i=1}^{r} I_{t_i,u_i}) \circ I_{M,X}$ is an isometric isomorphism

(iii) The following are equivalent

a) $C_o(M,X) \cong C_o(N,Y)$

b) $r = \tilde{r}$, and there is a permutation $\omega : \{1,\ldots,r\} \to \{1,\ldots,r\}$ such that $n_i M \cong m_{\omega(i)} N$ and $\tilde{X}_i \cong \tilde{Y}_{\omega(i)}$ for every $i \in \{1,\ldots,r\}$

Proof:

(i) We only have to apply lemma 11.11 to the isometric isomorphism
$\hat{I} := I_{N,Y} \circ I \circ I^{-1}_{M,X} : \prod_{i=1}^{r\,\infty} C_o(n_i M, \tilde{X}_i) \to \prod_{j=1}^{\tilde{r}\,\infty} C_o(m_j N, \tilde{Y}_j)$

(ii) This follows immediately from lemma 8.1.

(iii) "a ⇒ b": This is a consequence of (i).

"b ⇒ a": This is obvious. □

For the special case X = Y the permutation ω must be the identity (since $\tilde{X}_i = \tilde{X}_{\omega(i)}$ and the $\tilde{X}_i$ are pairwise not isometrically isomorphic). We thus obtain the following generalization of th. 7.1:

11.13 Theorem: Let X be an M-finite Banach space with canonical M-decomposition $X \cong \prod_{i=1}^{r} \tilde{X}_i^{n_i}$ and M, N nonvoid locally compact Hausdorff spaces. Then

(i) $C_o(M,X) \cong C_o(N,X)$ iff $(n_1 M \cong n_1 N \wedge n_2 M \cong n_2 N \wedge \ldots \wedge n_r M \cong n_r N)$

(ii) Every isometrical isomorphism $I: C_o(M,X) \to C_o(N,X)$ is of the form $I = I^{-1}_{N,X} \circ (\prod_{i=1}^{r} I_{t_i,u_i}) \circ I_{M,X}$, where the $t_i : n_i N \to n_i M$ are homeomorphisms and the $u_i : n_i N \to [X_i, X_i]_{iso}$ are continuous maps $(i=1,\ldots,r)$. □

Thus the question whether X has the Banach-Stone property is reduced to the topological problem whether $n_1 M \cong n_1 N \wedge \ldots \wedge n_r M \cong n_r N$ implies that $N \cong M$.

It is obvious that $\min_i n_i = 1$ is a sufficient condition, and we will see at once that this condition is also necessary.

In order to prove this we need the following result from the theory of Boolean algebras:

11.14 Theorem: For every countable commutative semigroup $(G,\circ)$ there is a family $(B_g)_{g\in G}$ of countable Boolean algebras such that

$$B_g \cong B_{g'} \text{ iff } g = g'$$

$$\text{and} \quad B_{g\circ g'} \cong B_g \times B_{g'} \qquad (\text{all } g,g' \in G),$$

i.e. $(G,\circ)$ admits a representation by isomorphism classes of Boolean algebras

Proof: We refer the reader to [69]. The proof is very technical and highly non-elementary.

Notes: 1. A much more elementary proof for a special case has been given in [56]: For every $n \in \mathbb{N}$, $n \geq 2$, there are non-isomorphic Boolean algebras A_n, B_n such that $(A_n)^n \cong (B_n)^n$; for the case $n = 2$ cf. also [87] .

2. The proof of th. 11.14 depends on the continuum hypothesis. Since we are interested only in a special case of this theorem (see the following corollary) we would like to know whether the assertion for the special case could be obtained in a more elementary way.

11.15 Corollary: There are compact Hausdorff spaces M_o, N_o such that $M_o \ncong N_o$ but $nM_o \cong nN_o$ for every $n \in \mathbb{N}$, $n \geq 2$.

Proof:

It is sufficient to construct a countable commutative semigroup $(G, \circ)$ with generators a,b such that $a \neq b$, $a^2 = b^2$, $a^3 = b^3$ (so that $a^n = b^n$ for every $n \geq 2$).

With M_o = the Stonean space of B_a and

N_o = the Stonean space of B_b (where $(B_g)_{g\in G}$ as in th. 11.14)

it follows that $M_o \ncong N_o$ (since $B_a \ncong B_b$) and $nM_o \cong nN_o$ (since $(B_a)^n \cong B_{(a^n)} \cong B_{(b^n)} \cong (B_b)^n$) for every $n \geq 2$.

We generate G by $G := \{a\} \cup \{b^n \mid n \in \mathbb{N}\}$, and " $\circ$ " is defined by the following multiplication table:

	a	$b^n_{(n\in\mathbb{N})}$
a	b^2	b^{n+1}
$b^m_{(m\in\mathbb{N})}$	b^{m+1}	b^{n+m}

It is not hard to prove that $(G,\circ)$ is a semigroup with the claimed properties. □

The following theorem is the main result of this section. It characterizes completely the M-finite Banach spaces with the (strong) Banach-Stone property.

11.16 Theorem: Let X be an M-finite Banach space with canonical M-decomposition $X \cong \prod_{i=1}^{r} {}^{\infty} \tilde{X}_i^{n_i}$. Then

(i) X has the Banach-Stone property iff $\min_i n_i = 1$

(ii) X has the strong Banach-Stone property iff $r = n_1 = 1$

(i.e. iff $Z(X) = \mathbb{K}\,\mathrm{Id}$)

Proof:

(i) It is clear from th. 11.13 that $\min n_i = 1$ implies that X has the Banach-Stone property. Conversely, if $\min n_i > 1$, then $n_i M_o \cong n_i N_o$ for $i=1,\ldots,r$ (M_o, N_o as in cor. 11.15), i.e. $C_o(M_o,X) \cong C_o(N_o,X)$ by th. 11.13. Since $M_o \not\approx N_o$ this implies that X does not have the Banach-Stone property.

(ii) If $r = n_1 = 1$, then $Z(X) = \mathbb{K}\,\mathrm{Id}$ so that X has the strong Banach-Stone property by th. 8.11. If $n_1 > 1$ or $r > 1$, then X contains a nontrivial M-summand. It has already been proved (cf. the note on p. 149) that X does not have the strong Banach-Stone property in this case. □

Applications

1. Let X be a two-dimensional Banach space. If $Z(X)$ is not one-dimensional, then $X \cong l_2^\infty$. It follows that there is essentially

one two-dimensional space (namely l_2^∞) which does not have the strong Banach-Stone property.

2. For three-dimensional spaces X there are three essentially different cases.

   case 1: $r = 1$, $n_1 = 1$

   In this case X has the strong Banach-Stone property

   case 2: $r = 1$, $n_1 = 3$

   This means that $X \cong l_3^\infty$; X does not have the Banach-Stone property

   case 3: $r = 2$, $n_1 = n_2 = 1$

   X has the Banach-Stone property but not the strong Banach-Stone property

   Note: This proves th. 8.7

3. Let Y be a Banach space such that $\dim Y > 1$ and $Z(Y) = \mathbb{K}\,Id$. Then $\mathbb{K} \times Y$ (provided with the supremum norm) has the Banach-Stone property.

   (This follows from th. 11.16 since $\mathbb{K} \times Y$ is M-finite with canonical M-decomposition $\mathbb{K}^1 \times Y^1$, i.e. we have $r = 2$, $n_1 = n_2 = 1$.)

   In particular, every cylindrical Banach space X such that $\dim X > 2$ has the Banach-Stone property (this is a consequence of cor.4.23).

The following proposition shows that in order to treat M-finite Banach spaces with the methods of section A it is sufficient to consider the $P^1_{X_o}$:

11.17 Proposition: Let X be an M-finite Banach space with canonical M-decomposition $X \cong \prod\limits_{i=1}^{r} {}^{\infty} \tilde{X}_i^{n_i}$. We identify X with its maximal function module representation as discussed in prop. 5.4(v):

base space $\quad K_X := \dot{\bigcup\limits_{i=1}^{r}} \Delta_i \quad (\Delta_i := \{1,\dots,n_i\})$

component spaces: $X_k := \tilde{X}_i$ for $k \in \Delta_i$.

Then, for every function module property P, we have

$$P(K_X,(X_k)_{k\in K_X},X) = \bigcup_{i\in\{1,\dots,r\}_P} P^1_{\tilde{X}_i}(K_X,(X_k)_{k\in K_X},X),\text{ where}$$

$\{1,\ldots,r\}_P$ is a subset of $\{1,\ldots,r\}$ ($P^1_{\tilde{X}_i}$ as on p. 184)

Proof: For $i\in\{1,\ldots,r\}$ and $\alpha,\beta\in\Delta_i$ the mapping $I_{i,\alpha,\beta}: X\to X$,

$$I_{i,\alpha,\beta}(x_1^1,\ldots,x_1^{n_1},\ \ldots\ ,x_i^1,\ldots,x_i^\alpha,\ldots,x_i^\beta,\ldots,x_i^{n_i},\ \ldots\ ,x_r^1,\ldots,x_r^{n_r})$$
$$:= (x_1^1,\ldots,x_1^{n_1},\ \ldots\ ,x_i^1,\ldots,x_i^\beta,\ldots,x_i^\alpha,\ldots,x_i^{n_i},\ \ldots\ ,x_r^1,\ldots,x_r^{n_r}),$$

is an isometrical isomorphism, and the associated map $t_{i,\alpha,\beta}$ (cf. cor. 4.17) maps α to β, β to α, and leaves all other points of K_X fixed. By the definition of function module properties we have $t_{i,\alpha,\beta}(P(K_X,(X_k)_{k\in K_X},X)) = P(K_X,(X_k)_{k\in K_X},X)$ so that $P(K_X,(X_k)_{k\in K_X},X)$ contains Δ_i if $P(K_X,(X_k)_{k\in K_X},X)\cap\Delta_i\neq\emptyset$.

Hence $\quad P(K_X,(X_k)_{k\in K_X},X) = \bigcup\{\Delta_i \mid P(K_X,(X_k)_{k\in K_X},X)\cap\Delta_i\neq\emptyset\}$

$\qquad\qquad = \bigcup\{P^1_{\tilde{X}_i}(K_X,(X_k)_{k\in K_X},X) \mid P(K_X,(X_k),X)\cap\Delta_i\neq\emptyset\}.$ □

11.18 Corollary: Let X, $(K_X,(X_k)_{k\in K_X},X)$, $\Delta_1,\ldots,\Delta_r$ be as in prop. 11.17. Then $\mathfrak{B}(K_X,(X_k)_{k\in K_X},X)$ is the Boolean algebra generated by the Δ_i, i.e. $\mathfrak{B}(K_X,(X_k)_{k\in K_X},X) = \{\bigcup_{i\in S}\Delta_i \mid S\subset\{1,\ldots,r\}\}$ □

Finally we generalize the result that, if $Z(X) = \mathbb{K}\,Id$ and $Z(Y) = \mathbb{K}\,Id$, $C_o(M,X)\cong C_o(N,Y)$ does not only imply that $M\cong N$ but also that $X\cong Y$ (th. 8.10). We will prove that this is true for all "square-free" M-finite Banach spaces.

11.19 Proposition: Let X and Y be M-finite Banach spaces with canonical M-decompositions $X\cong\prod_{i=1}^{r}{}^\infty\tilde{X}_i^{n_i}$ and $Y\cong\prod_{j=1}^{\tilde{r}}{}^\infty\tilde{Y}_j^{m_j}$, respectively. Suppose that $n_1=\ldots=n_r=m_1=\ldots=m_{\tilde{r}}=1$ and that there are nonvoid locally compact Hausdorff spaces M, N such that $C_o(M,X)\cong C_o(N,Y)$. Then X and Y are isometrically isomorphic

Proof: It follows from lemma 11.11 that $r=\tilde{r}$ and that $\tilde{X}_i\cong\tilde{Y}_{\omega(i)}$ for $i=1,\ldots,r$, where $\omega:\{1,\ldots,r\}\to\{1,\ldots,r\}$ is a suitable permutation. Hence $X\cong\prod_{i=1}^{r}{}^\infty\tilde{X}_i\cong\prod_{i=1}^{r}{}^\infty\tilde{Y}_{\omega(i)}\cong\prod_{i=1}^{r}{}^\infty\tilde{Y}_i\cong Y$. □

C. Generalizations of the Banach-Stone theorem: S U M M A R Y

Let X be a fixed Banach space. How can the results of these notes be applied in order to decide whether or not X has the Banach-Stone property or the strong Banach-Stone property ?

First it is important to determine $Z(X)$

[If X has an additional (order or algebraic) structure, then $Z(X)$ can be obtained by using the results of section C and section D in chapter 5 (cf. also the references in chapter 6).

In the general case $Z(X)$ can (possibly) be constructed by using, for example,

th. 3.13 (characterization of the operators in $Z(X)$)

prop. 5.1, prop. 8.12 (sufficient conditions such that $Z(X) = \mathbb{K}\,\mathrm{Id}$)

or th. 5.9 ($Z(X')$ can be constructed from $C(X)$) .]

First case: Suppose that it can be shown that $Z(X)$ is trivial. Then X has the strong Banach-Stone property, i.e. X behaves exactly as the scalar field in the classical version of the Banach-Stone theorem (th. 8.11).

Second case: If $Z(X)$ is finite-dimensional (but not necessarily one-dimensional) then the results of the preceding section can be applied (we recall that, for example, every reflexive space has this property).

It is important to know the M-exponents $n_1,\ldots,n_r$ of X which can easily be obtained from the minimal nonzero M-summands of X (see chapter 5, section B).

We have proved that X has the Banach-Stone property iff $\min n_i = 1$

(i.e. iff there is a minimal nonzero M-summand which is not isometrically isomorphic to any other M-summand of X ; th. 11.16).

Third case: Suppose that Z(X) is not finite-dimensional. In this case there are no results available which guarantee that X has the strong Banach-Stone property. However, if X has the local cns property, then the results of section A (of this chapter) can be applied, i.e. we can investigate whether X has the Banach-Stone property.

Since only very pathological examples of Banach spaces failing to have the local cns property are known (see chapter 9) there is a good chance that X is not such an exceptional space.

How can one prove that X has the local cns property ? If it is not possible to show directly that X has a cns (e.g. by investigating the examples and the results in the first part of chapter 9) and thus the local cns property it will be necessary to construct a maximal function module representation of X.

There are essentially two ways of obtaining such a representation. The first way is to apply th. 4.14, the second is to consider a "natural" function module representation of X and to prove that this is in fact a maximal function module representation (cf. p. 99).

Suppose that a maximal function module representation $(K_X,(X_k)_{k\in K_X},X)$ of X has been constructed and that it can be shown that X has the local cns property. Then, for every hereditary function module property P there is a subset $P(K_X,(X_k)_{k\in K_X},X)$ of the base space such that $C_o(M,X) \cong C_o(N,X)$ implies that $M \times P(K_X,(X_k)_{k\in K_X},X) \cong N \times P(K_X,(X_k)_{k\in K_X},X)$.

In view of our problem we need a P such that $P(K_X,(X_k)_{k\in K_X},X)$ contains exactly one element. This means that we have to look carefully for a topological and/or functional analytical property of function modules which is satisfied at only one point of K_X. A number of candidates for P have been considered in section A of this chapter.

There seems to be no way of modifying our results in such a way that arbitrary Banach spaces can be discussed. However, the local cns property of X is only essential for the proof of th. 10.10 so that if it is possible to obtain information concerning maximal function module representations of $C_o(M,X)$ by any other construction (for all locally compact Hausdorff spaces M), then function module properties can be applied with obvious modifications even if X fails to have the local cns property.

12. Remarks

The results of part II (apart from th. 7.1, th. 8.3, th. 8.4, th.8.5, th.8.6, th.8.7, th. 8.9, and th. 11.14) are due to the author; some of the results in section A of chapter 11 have been obtained together with U. Schmidt-Bichler .

The essential features of most of our theorems have already been published in a number of papers (cf. [10] , [11] , [17]).

The unified treatment of

- arbitrary (not necessarily real) Banach spaces
- arbitrary locally compact (not necessarily compact) Hausdorff spaces
- arbitrary Banach spaces with the local cns property (which do not necessarily have a cns)

as well as some results in chapter 9 and chapter 10 are new.

In this chapter we will discuss a generalization of the technique of function module properties (section A) and a supplement concerning square Banach spaces with a cns (section B). Finally, in section C, we indicate some open problems.

A. The Banach-Stone property for $\mathfrak{C}$

By definition, a Banach space X has the Banach-Stone property if $C_o(M,X) \cong C_o(N,X)$ implies that $M \cong N$ for all locally compact Hausdorff spaces M, N. It is not hard to modify our methods in such a way that arbitrary classes of locally compact Hausdorff spaces can be treated:

12.1 Definition: Let $\mathfrak{C}$ be a class of nonvoid locally compact Hausdorff spaces. A Banach space X is said to have the Banach-Stone property for $\mathfrak{C}$ if $C_o(M,X) \cong C_o(N,X)$ implies that $M \cong N$ whenever

M, N are in $\mathfrak{C}$.

It is obvious how $\mathfrak{C}$-hereditary function module properties have to be defined. Clearly there are in general many more $\mathfrak{C}$-hereditary than hereditary function module properties.

(For example, if

$\mathfrak{C}_{\text{connected}}$:= the class of nonvoid connected locally compact Hausdorff spaces

$\mathfrak{C}_{\text{metrizable}}$:= the class of nonvoid metrizable locally compact Hausdorff spaces ,

then $P^6_{\text{connected}}$ is $\mathfrak{C}_{\text{connected}}$-hereditary and $P^2_{\aleph_o}$ is $\mathfrak{C}_{\text{metrizable}}$-hereditary, but neither $P^6_{\text{connected}}$ nor $P^2_{\aleph_o}$ is hereditary; for definitions, see p. 184 and p. 185.)

As in th. 11.9 we then have:

Let X be a Banach space with the local cns property (as usual we assume that X is maximally represented as a function module $(K_X,(X_k)_{k\in K_X},X)$). If there exists a $\mathfrak{C}$-hereditary function module property P such that $P(K_X,(X_k)_{k\in K_X},X)$ contains exactly one element, then X has the Banach-Stone property for $\mathfrak{C}$.

Note: For M, N $\in \mathfrak{C}$ we have $M\times P(K_X,(X_k)_{k\in K_X},X) \cong N\times P(K_X,(X_k)_{k\in K_X},X)$ for every $\mathfrak{C}$-hereditary function module property P (provided that $C_o(M,X) \cong C_o(N,X)$). Even if there is no $\mathfrak{C}$-hereditary P such that card $P(K_X,(X_k)_{k\in K_X},X) = 1$ it is sometimes possible to conclude that $M \cong N$ (i.e. that X has the Banach-Stone property for $\mathfrak{C}$).

For example, it is obvious that $nM \cong nN$ implies that $M \cong N$ for connected spaces M, N so that every M-finite Banach space has the Banach-Stone property for $\mathfrak{C}_{\text{connected}}$ (more generally: every Banach space X for which there exists a $\mathfrak{C}$-hereditary function module property P such that $P(K_X,(X_k)_{k\in K_X},X)$ is finite).

B. Square Banach spaces with a centralizer-norming system

In this section we will show that square Banach spaces with a cns have a particularly simple form.

Let $(K,(X_k)_{k\in K},X)$ be a function module such that $\dim X_k \leq 1$ for every $k \in K$ and $Z(X) = \{M_h \mid h \in CK\}$. We recall that every square Banach space (cf. p. 132) has this form.

12.2 Definition: ($(K,(X_k)_{k\in K},X)$ as above)

(i) We say that X is globally trivial if there is an $x_o \in X$ such that $X = \{ hx_o \mid h \in CK\}$

(i.e. if X is a renorming of the space CK)

(ii) For $k_o \in K$ we say that X is locally trivial at k_o if there is a compact neighbourhood U of k_o such that $X|_U$ is globally trivial (i.e. if there is an $x_o \in X$ such that $\{x|_U \mid x \in X\} = \{h(x_o|_U) \mid h \in CK\}$)

Examples/Remarks:

1. C_oM (M a locally compact space) is globally trivial iff M is compact. If C_oM is maximally represented as on p. 88 (example 2, with $K = \beta M$), then C_oM is locally trivial at every point of M but not at the points of $\beta M \setminus M$.
2. If K is a compact Hausdorff space and $x_o : K \to \mathbb{K}$ a function such that $k \mapsto |x_o(k)|$ is upper semicontinuous and $0 < \inf_k |x_o(k)|$, then $X := \{hx_o \mid h \in CK\}$ (provided with the supremum norm) is a globally trivial square Banach space.
3. A space which is locally trivial at every point of the base space is not necessarily globally trivial (consider, for example, the real Banach space X_s with $-1 < s < 0$; cf. p. 100).

Example 2 describes the most general globally trivial space as the following lemma shows:

12.3 Lemma: Let X be a globally trivial square Banach space and

x_o as in def. 12.2(i). Then $k \mapsto |x_o(k)|$ is upper semicontinuous and $\inf_k |x_o(k)| > 0$.

Proof: $|x_o(\cdot)|$ is upper semicontinuous since X is a function module. Suppose that $\inf |x_o(k)| = 0$. We will show that this is in contradiction to the completeness of X.

Firstly we show that there is a disjoint sequence $U_1, U_2, \ldots$ of nonvoid open subsets of K such that $\|x_o|_{U_n}\| \leq 1/n^2$ for $n \in \mathbb{N}$.

The sets $O_n := \{k \mid |x_o(k)| < 1/n^2\}$ are open, nonvoid and decreasing. We consider the sequence $O_1^- \supset O_2^- \supset O_3^- \supset \ldots$.

First case: There are indices $n_1, n_2, n_3, \ldots$ such that $O_{n_1}^- \supsetneq O_{n_2}^- \supsetneq O_{n_3}^- \supsetneq \ldots$

In this case we define
$$U_1 := O_{n_1} \setminus O_{n_2}^-$$
$$U_2 := O_{n_2} \setminus O_{n_3}^-$$
$$\ldots\ldots$$

Second case: There is an index n_o such that $O_{n_o}^- = O_{n_o+1}^- = \ldots$.

Then $O_{n_o}, O_{n_o+1}, \ldots$ is a decreasing sequence of open dense subsets of $O_{n_o}^-$ so that, since compact spaces are Baire spaces, $\{k \mid x_o(k)=0\} = \bigcap_{n \geq n_o} O_n$ is dense in O_{n_o}. We choose points $k_1, k_2, \ldots \in O_{n_o}$ and open neighbourhoods $\tilde{U}_n$ of k_n (which are contained in O_{n_o}) such that $x_o(k_n) = 0$, $\tilde{U}_n \cap \tilde{U}_m = \emptyset$ for $n, m \in \mathbb{N}$, $n \neq m$. Then the $U_n := \tilde{U}_n \cap O_n$ have the claimed properties.

Choose functions $h_1, h_2, \ldots \in CK$ such that $\|h_n\| = 1$ and $h_n|_{K \setminus U_n} = 0$ (all $n \in \mathbb{N}$). The pointwise defined function $\alpha := (\sum_{n=1}^{\infty} nh_n)x_o$ is contained in $\{hx_o \mid h \in CK\}^-$ but not in $\{hx_o \mid h \in CK\}$ in contradiction to the completeness of X (note that, for $h \in CK$, $\alpha = hx_o$ would imply that $h|_{U_n} = nh_n|_{U_n}$ since $\{k \mid x_o(k) \neq 0\}$ is dense in U_n and h and nh_n are continuous; this is not possible since $\|h\| < \infty$). □

By the following proposition, local and global triviality can be described by means of (local) centralizer-norming systems:

12.4 Proposition: Let $(K, (X_k)_{k \in K}, X)$ be as above and $k_o \in K$.

(i) X is globally trivial iff X contains a cns consisting of a single element

(ii) The following are equivalent:

a) X is locally trivial at k_o

b) k_o has a local cns

c) k_o has a local cns consisting of a single element

(iii) The following are equivalent:

a) X is locally trivial at every point of K

b) X has a cns

c) X is a finitely generated CK-module (i.e. there are $x_1,\dots,x_n \in X$ such that $X = \text{lin}\{hx_i \mid i\in\{1,\dots,n\},\ h\in CK\}$)

<u>Proof</u>:

(i) If X is globally trivial and x_o as in def. 12.2(i), then $\{x_o\}$ is a cns by lemma 12.3 and prop. 9.4.

Conversely, suppose that $\{x_o\}$ is a cns in X. We have to show that $X = \{hx_o \mid h\in CK\}$, i.e. $x/x_o \in CK$ for every $x\in X$ (note that $r := \inf_k |x_o(k)| > 0$ by prop. 9.4 so that $1/x_o$ is defined).

Let $x\in X$ and $k_o\in K$ be arbitrary. By prop. 9.4 $x_o(k_o)$ is different from 0 so that there is an $a\in\mathbb{K}$ such that $(ax_o - x)(k_o) = 0$.

Since $ax_o - x\in X$ it follows that, for $\varepsilon > 0$, there is a neighbourhood U_ε of k_o such that $|(ax_o - x)(k)| < \varepsilon$ for $k\in U_\varepsilon$. This implies that $|(x/x_o)(k) - (x/x_o)(k_o)| < \varepsilon/(|x_o(k)|) \leq \varepsilon/r$ which proves that x/x_o is continuous at k_o.

(ii) We have "a $\Leftrightarrow$c" by (i), and "c $\Rightarrow$ b" is trivially satisfied. It remains to show that b implies c.

Let $x_1,\dots,x_n$ be a local cns for k_o. By definition, there is a neighbourhood U of k_o such that $r := \inf_{k\in U} \max_i |x_i(k)| > 0$. In particular there is an index i_o such that $|x_{i_o}(k_o)| \geq r$.

The functions $(x_i(k_o)/x_{i_o}(k_o))x_{i_o} - x_i$ vanish at k_o and are contained in X so that there is a neighbourhood $\tilde{U}$ of k_o such that $|(x_i(k_o)/x_{i_o}(k_o))x_{i_o}(k) - x_i(k)| < r/2$ for $k\in\tilde{U}$ and $i=1,\dots,n$. It follows that $|x_{i_o}(k)| \geq r/2(\max_i |x_i(k_o)/x_{i_o}(k_o)|)^{-1}$ for $k\in U\cap\tilde{U}$, i.e.

$\{x_{i_o}\}$ is a local cns for k_o.

(iii) "a ⇔ b" follows from (ii) and the fact that a Banach space X has a cns iff every point of the base space (of a maximal function module representation) has a local cns.

"b ⇒ c": Let $x_1,\ldots,x_n$ be a cns for X and $x \in X$. For $k \in K$ we may choose an $i_k \in \{1,\ldots,n\}$ and a neighbourhood U_k of k such that $\{x_{i_k}\}$ is a local cns for k (i.e. $\inf_{l \in U_k} |x_{i_k}(l)| > 0$; cf. the proof of (ii), b ⇒ c). By (ii)a ⇔ c we have $x|_{U_k} \in \{hx_{i_k}|_{U_k} \mid h \in CK\}$ so that $x|_{U_k} = h_k^* x_{i_k}|_{U_k}$ for a suitable function $h_k \in CK$.

Let $U_{k_1},\ldots,U_{k_m}$ be a finite covering of K. We then have $x = \sum_{j=1}^{m} h_j h_{k_j}^* x_{k_j}$, where $h_1,\ldots,h_m$ is a partition of unity subordinate to $U_{k_1},\ldots,U_{k_m}$. This proves c.

"c ⇒ b": It has to be shown that $\inf_k \max_i |x_i(k)| > 0$ (with $x_1,\ldots,x_n$ as in c).

Suppose that $\inf_k \max_i |x_i(k)| = 0$. We construct a disjoint sequence of nonvoid open subsets of K such that $\max_i \|x_i|_{U_m}\| < 1/m^2$ and functions $h_1,h_2,\ldots \in CK$ such that $\|h_m\| = 1$ and $h_m|_{K \setminus U_m} = 0$ similarly to the proof of lemma 12.3.

For every $m \in \mathbb{N}$ there is an $i_m \in \{1,\ldots,n\}$ such that $\|x_{i_m}|_{\tilde{U}_m}\| = \max_i \|x_i|_{\tilde{U}_m}\|$ (where $\tilde{U}_m := \{k \mid |h_m(k)| \geq 1/2\}$). We choose $i_o \in \{1,\ldots,n\}$ such that $i_m = i_o$ for infinitely many indices, say $m_1,m_2,\ldots$.

The (pointwise defined) function $\alpha := (\sum_{m=1}^{\infty} mh_m)x_{i_o}$ is obviously contained in $\{hx_{i_o} \mid h \in CK\}^- \subset \overline{\text{lin}}\,\{hx_i \mid h \in CK,\ i=1,\ldots,n\}$. We claim that it is not contained in $\text{lin}\{hx_i \mid h \in CK,\ i=1,\ldots,n\}$ (in contradiction to the completeness of X).

Suppose that $\alpha = \sum_{i=1}^{n} h_i^* x_i$ for suitable $h_i^* \in CK$. In particular we have $m_j h_{m_j} x_{i_o}|_{\tilde{U}_{m_j}} = \sum_{i=1}^{n} h_i^* x_i|_{\tilde{U}_{m_j}}$ for $j=1,2,\ldots$ so that, since $\|x_{i_o}|_{\tilde{U}_{m_j}}\| = \max_i \|x_i|_{\tilde{U}_{m_j}}\|$ and $|h_{m_j}| \geq 1/2$ on U_{m_j}, there must be an $i_j \in \{1,\ldots,n\}$ such that $\|h_{i_j}^*\| \geq m_j/2n$ (note that $\|x_{i_o}|_{\tilde{U}_{m_j}}\| > 0$ since $\{k \mid X_k \neq \{0\}\}^-$

= K).

This is not possible since $\max_i \|h_i^*\| < \infty$. □

C. Problems

Problem 1: Does theorem 11.9 characterize Banach spaces with the Banach-Stone property ?

More precisely: If X is a Banach space with the local cns property and if X has the Banach-Stone property, is it possible to define a hereditary function module property P such that $P(K,(X_k)_{k\in K},X)$ contains exactly one element ?

Problem 2: Prop. 11.17 describes all hereditary function module properties which are essential in order to treat M-finite Banach spaces.

Are there other/larger classes of Banach spaces for which a similar result can be proved (e.g. CK-spaces, spaces with a cns, ...) ?

Problem 3: Is there a compact Hausdorff space K which contains more than one point such that CK has the Banach-Stone property ?

By cor. 7.3 this is equivalent to the following topological problem:

> If K is a compact Hausdorff space such that card $K > 1$, are there nonhomeomorphic (locally) compact Hausdorff spaces M, N such that $M \times K \cong N \times K$?

(That this is the case for $K = \{1,...,n\}$, $n \geq 2$, is just the assertion of cor. 11.15; the cases $K = [0,1]$ and $K = \beta\mathbb{N}$ have been discussed on p. 143).

Notation index

Functions:	$\underline{x}$ means the function which assumes the value x at every point
Isomorphisms:	$\cong$ (see p.1, p.4)
Operations in a Boolean algebra:	$\wedge, \vee, ^\wedge$ (see p.5)
Order structures:	$\leq$ (see p. 5, 19, 42, 64)
Topology:	M^- means the closure, M^o the interior of a subset M in a topological space

A	a B^*-algebra
αL	the one-point compactification of a space L, 4
AK	the space of continuous affine functions on K, 154
A_{sa}	the real Banach space of self-adjoint elements in A
a_T	the eigenvalue-function of a multiplier, 54
$\mathfrak{A}$	a Boolean algebra, 5
βL	the Stone-Čech compactification of a space, 4
B(X)	the space of continuous linear operators on X, 2
B(x,r)	the open ball with centre x and radius r, 45
c, c_o	the spaces of convergent sequences and null sequences, 1
$\mathbb{C}$	the complex scalar field
$\mathfrak{C}$	a class of locally compact Hausdorff spaces, 201
$CM, C_{\mathbb{C}}M, C_{\mathbb{R}}M, C^bM$	spaces of continuous scalar-valued functions, 1
$C_o(M,X)$	the space of X-valued continuous functions which vanish at infinity, 1
$C^\ell(X)$	the Cunningham algebra of X, 29
$C_\infty(X)$	the Cunningham-∞-algebra, 31
$C_{k,\Delta}$	the T-sets in a function module, 149
co; $\overline{co}$	convex hull, closed convex hull
card M	the cardinality of M
cns	centralizer-norming system, 152
δ_k	evaluation functional, 4, 55
D(x,r)	the closed ball with centre x and radius r, 46
d(x,J)	the distance from a point x to a subset J, 85
E_X	the extreme points of the unit ball of X', 54
I	an isometrical isomorphism
$I_{t,u}$	an isometrical isomorphism defined by means of u and t, 141
I_ω	a permutation operator, 190
$I_{M,X}$	a canonical isometrical isomorphism, 191
Im	the imaginary part of a scalar
Id	the identity operator
J^π	the annihilator of J in X', 3
$J^\perp$	the complementary summand, 9
$J_{\mathfrak{C}}$	M-summand in a CK-space, 10, 36
$J^{(\perp)}$	the natural candidate for the complementary L-summand, 42
$\mathbb{K}$	the field of real or complex numbers, 1
K_X	base space of the maximal function module representation, 63

K_X^*	a subset of K_X, 103
K_M	the base space of the associated representation, 172
$K(H)$	the space of compact operators on H, 11
$K(x,r)$	a subset of $X' \times \mathbb{R}$, defined by $x \in X$ and $r \geq 0$, 44
l^1	the space of absolutely convergent sequences, 1
l_n^∞	$\mathbb{K}^n$, provided with the supremum norm, 1
lin, $\overline{\text{lin}}$	linear span, closed linear span of a set
local cns	local centralizer-norming system, 160
m	the space of bounded sequences, 1
M_h	multiplication operator associated with h, 41, 84
$m(E_X, \mathbb{K})$	the space of bounded functions from E_X to $\mathbb{K}$, 54
Mult(X)	the space of multipliers on X, 54
nK	the disjoint union of n copies of K, 4
non P	a function module property, 185
ω_X	a canonical isomorphism, 29
Ω_X	the Stonean space of the algebra of L-projections, 29
Ω_X^∞	the Stonean space of the algebra of M-projections, 31
$\mathbb{P}_M, \mathbb{P}_L, \mathbb{P}_L(X), \mathbb{P}_M(X)$	the set of all L- or M-projections, 12
$\mathbb{P}_p(X)$	the set of all L^p-projections, 123
$P_J(x)$	the set of best approximation to x in J, 126
$P, P_{X_o}^1, P_c^2$	function module properties, 183, 184
$P(K,(X_k),X)$	the subset of K defined by P, 183
$P^-, P^o, \bigwedge P_i, \bigvee P_i$	function module properties, 185
$\mathfrak{P}(K,(X_k),X)$	a Boolean algebra of subsets of K, 187
$\mathfrak{P}_{her}(\ldots)$	a Boolean algebra of subsets of K, 187
$\Pi^1 X_i, \Pi^\infty X_i$	products of Banach spaces, 17, 75
ΠT_k	operator product, 83
Prim(X)	the collection of primitive M-ideals, 73
$\mathbb{R}$	the real number field
R	a function module representation, 90
$R_2 \lesssim R_1$	R_1 is finer than R_2, 94
$R_2 \approx R_1$	R_1 and R_2 are equivalent, 95
$R_\lambda^2(x)$	a certain intersection of balls, 56
Re	the real part of a scalar
T'	the transposed operator, 3
T^*	the adjoint operator, 62
$\mathfrak{T}_{ex}$	a topology on B(X), 88
X	a Banach space
$X_\mathbb{R}$	the underlying real space
[X,Y]	the space of operators from X to Y, 1
X'	the dual space of X, 2
$[X,Y]_{iso}$	the set of isometric isomorphisms from X to Y, 2, 141
$X\|_L$	restriction of a function module, 79
X_s	a space of continuous functions, 100
X_M	the associated function module, 172
Z(X)	the centralizer of X
$Z_\rho(X)$	a subalgebra of the centralizer, 90

S u b j e c t   i n d e x

adjoint operator 62
Alaoglu-Bourbaki theorem 2
annihilator (of a subspace) 3
associated function module representation 172

Banach-Stone property, strong Banach-Stone property 142
Banach-Stone property for $\mathfrak{C}$ 201
Banach-Stone theorem 138
base space (of a function module) 77
Boolean algebra 5

canonical M-decomposition 112
carrier projection 118
centralizer 62
centralizer-norming system (cns) 152
(local) centralizer-norming system 160
$\mathfrak{C}$-hereditary function module property 202
clopen = closed and open
complementary L-summand, M-summand 11
complete Boolean algebra 6
component spaces of a function module 77
Cunningham algebra 29
cylindrical Banach space 146

Dauns-Hofmann theorem 74
(generalized version of the) Dauns-Hofmann theorem 60
discrepant T-sets 144

equivalent function module representations 95
extremally disconnected 4
extreme functional =
extreme point in the unit ball of X'

face(x) 41
finer 94
function module 77
function module property 183
(hereditary) function module property 186
function module representation 90

globally trivial 203

Hahn-Banach theorem 2
hereditary function module representation 186
hull 73
hyperstonean 4

increasing family 20
induced order 5

Krein-Milman theorem 2
Krein-Šmulian theorem 2

Lindenstrauß space 28
L-M-theorem 23
local centralizer-norming system 160
local cns property 160

locally trivial 203
L^p-projection 122
L-projection 12
L^p-summand 122
L-summand 9

maximal function module representation 99
M-bounded operator 55
(canonical) M-decomposition 112
M-exponent 112
M-finite 110
M-ideal 34
multiplier 54
Mult(X) 54
M-projection 12
M-summand 9

n-ball property for open/closed balls 46
norm resolution 92
(= the function $\|x(\cdot)\|$, x an element of a function module)

operator = continuous linear map 1
order continuous 4

partition of unity 5
primitive M-ideal 73
projection = idempotent operator

$\mathbb{R}$-determined 22
restriction of a function module 79

selfadjoint subalgebra 4
square Banach space 132
Stonean space 6
strong Banach-Stone property 142
strong operator topology 2
structurally continuous
(=continuous with respect to the structure topology)
structure topology 58,73

transposed operator 3
trivial centralizer 63
trivial L-projection/M-projection 12
trivial Lsummand/M-summand 10
trivial M-ideal 36
T-set 139

weak operator topology 2
weak*-topology 2

References

[1] P.Alexandroff-H.Hopf: Topologie
Chelsea Publ. Comp., New York 1972

[2] E.M.Alfsen: Compact convex sets and boundary integrals
Springer Verlag, Erg.d.Math. 57, Berlin 1971

[3] E.M.Alfsen-E.G.Effros: Structure in real Banach spaces I
Ann. of Math. 96, 1972, 98-128

[4] E.M.Alfsen-E.G.Effros: Structure in real Banach spaces II
Ann. of Math. 96, 1972, 129-173

[5] D.Amir: On isomorphisms of continuous function spaces
Israel J. of Math. 3, 1965, 205-210

[6] W.G.Badé: On Boolean algebras of projections and algebras of operators
Trans. of the Am. Math. Soc. 80, 1955, 345-360

[7] S.Banach: Théorie des opérations linéaires
Warschau, 1932

[8] E.Behrends: L^p-Struktur in Banachräumen
Studia Math. 55, 1976, 71-85

[9] E.Behrends: L^p-Struktur in Banachräumen II
Studia Math. 62, 1977, 47-63

[10] E.Behrends: An application of M-structure to theorems of the Banach-Stone type
in: Notas de Matemática, North Holland Math. Studies 27 (Proceedings on the Paderborn Conference on Functional Analysis 1976), 1977, 29-49

[11] E.Behrends: On the Banach-Stone theorem
Math. Annalen 233, 1978, 261-272

[12] E.Behrends: T-sets in function modules and an application to theorems of the Banach-Stone type
Revue Roumaine de Math. pures et appl. (to appear)

[13] E.Behrends: The centralizer of tensor products of Banach spaces (a function space representation)
Pac. J. of Math. (to appear)

[14] E.Behrends: M-structure in tensor products of Banach spaces
in: Notas de Matemática, North Holland Math. Studies (Proceedings of the Rio de Janeiro Conferece on Functional Analysis, Approximation theory, and Complex Analysis 1978) (to appear)

[15] E.Behrends: Norm interval respecting operators
Studia Math. (to appear)

[16] E.Behrends et al.: L^p-structure in real Banach spaces
Lect. Notes in Math. 613, Springer Verlag, Berlin 1977

[17] E.Behrends-U.Schmidt-Bichler: M-structure and the Banach-Stone theorem
Studia Math. (to appear)

[18] D.G.Bourgin: Approximately isometric and multiplicative transformations on continuous function rings
Duke Math. J. 16,1949, 385-397

[19] R.A.Bowshell: Continuous associative sums of Banach spaces
(Preprint)

[20] M.Cambern: A generalized Banach-Stone theorem
Proc. of the Am. Math. Soc. 17, 1966, 396-400

[21] M.Cambern: On isomorphisms with small bounds
Proc. of the Am. Math. Soc. 18, 1967, 1062-1066

[22] M.Cambern: Isomorphisms of C_oY onto C_oX
Pac. J. of Math. 35, 1970, 307-312

[23] M.Cambern: On mappings of spaces of functions with values in a Banach space
Duke Math. J. 42, 1975, 91-98

[24] M.Cambern: Isomorphisms of spaces of continuous vector-valued functions
Ill. J. of Math. 20, 1976, 1-11

[25] M.Cambern: The Banach-Stone property and the weak Banach-Stone property in three-dimensional spaces
Proc. of the Am. Math. Soc. 67, 1977, 55-61

[26] M.Cambern: Reflexive spaces with the Banach-Stone property
Revue Roumaine de Math. pures et appl. (to appear)

[27] C.K.Chui et al.: L-ideals and numerical range preservation
Ill. J. of Math. 21, 1977, 365-373

[28] H.B.Cohen: A second dual method for C(X)-isomorphisms
J. of Funct. Anal. 23, 1976, 107-118

[29] F.Cunningham: L-structure in L-spaces
Trans. of the Am. Math. Soc. 95, 1960, 274-299

[30] F.Cunningham: M-structure in Banach spaces
Proc. of the Cambr. Phil. Soc. 63, 1967,613-629

[31] F.Cunningham: Square Banach spaces
Proc. of the Cambr. Phil. Soc. 66, 1969,553-558

[32] F.Cunningham-E.G.Effros-N.M.Roy: M-structure in dual Banach spaces
Israel J. of Math. 14, 1973, 304-308

[33] F.Cunningham-N.M.Roy: Extreme functionals on an upper semicontinuous function space
Proc. of the Am. Math. Soc. 42, 1974, 461-465

[34] R.Danckwerts-S.Göbel-K.Meyfarth: Über die Cunningham-∞-Algebra und den Zentralisator reeller Banachräume
Math. Annalen 220, 1976, 163-169

[35] J.Dauns-K.H.Hofmann: Representation of rings by sections
Memoirs of the Am. Math. Soc. 83, 1968

[36] J.Diestel: Geometry of Banach spaces - selected topics
Lect. Notes in Math. 485, Springer Verlag, Berlin 1975

[37] J.Dixmier: Les fonctionnelles linéaires sur l'ensemble des opérateurs bornés d'un espace de Hilbert
Ann. of Math. 51, 1950, 387-408

[38] J.Dixmier-A.Douady: Champs continues d'espaces hilbertien et de C^*-algèbres
Bull. Soc. math. France 91, 1963, 227-284

[39] N.Dunford-J.T.Schwartz: Linear operators I
Interscience publishers, New York 1971

[40] N.Dunford-J.T.Schwartz: Linear operators III
Interscience publishers, New York 1971

[41] S.Eilenberg: Banach space methods in topology
Ann. of Math. 43, 1942, 568-579

[42] G.A.Elliott: An abstract Dauns-Hofmann-Kaplansky multiplier theorem
Canadian J. of Math. 27, 1975, 827-836

[43] G.A.Elliott-D.Olesen: A simple proof of the Dauns-Hofmann theorem
Math. Scand. 34, 1974, 231-234

[44] R.Evans: Projektionen mit Normbedingungen in reellen Banachräumen
Dissertation, Freie Universität Berlin, 1974

[45] R.Evans: A characterization of M-summands
Proc. of the Cambr. Phil. Soc. 76, 1974, 157-159

[46] R.Evans: Embedding C(K) in B(X)
(Preprint)

[47] H.Fakhoury: Existence d'une projection continue de meilleure approximation dans certaines espaces de Banach
J. Math.pures et appl. 53, 1974, 1-16

[48] H.Fakhoury: Sur les M-ideaux dans les espaces d'opérateurs
(Preprint)

[49] R.H.Fox: On a problem of Ulam concerning cartesian products
Fundamenta math. 34, 1947, 271-287

[50] G.Gierz: Darstellung von Banachverbänden durch Schnitte in Bündeln
Mitteilungen aus dem Math. Sem. d. Univ. Gießen 125, Gießen 1977

[51] G.Gierz: Representation of spaces of compact operators and applications to the approximation property
Archiv d. Math. 30, 1978, 622-628

[52] R.Godement: Sur la théorie des représentations unitaires
Ann. of Math. 53, 1951, 68-124

[53] Y.Gordon: On the distance coefficient between isomorphic function spaces
Israel J. of Math. 8, 1970, 391-397

[54] P.Greim: Integraldarstellung von Banachräumen und Dualität
Dissertation, Freie Universität Berlin, 1975

[55] P.R.Halmos: Lectures on Boolean algebras
Springer Verlag, Berlin 1974

[56] W.Hanff: On some fundamental problems concerning isomorphism of Boolean algebras
Math. Scand. 5, 1957, 205-217

[57] J.Hennefeld: A decomposition for $B(X)^*$ and unique Hahn-Banach extensions
Pac. J. of Math. 46, 1973, 197-199

[58] B.Hirsberg: M-ideals in complex function spaces and algebras
Israel J. of Math. 12, 1972, 133-146

[59] K.Hoffman: Banach spaces of analytic functions
Prentice Hall Inc., Englewood Cliffs N.J., 1962

[60] K.H.Hofmann-K.Keimel: Sheaf theoretical concepts in analysis: Bundles and sheaves of Banach spaces, Banach C(X)-modules
TH Darmstadt, Preprint 394, 1978

[61] R.Holmes: M-ideals in approximation theory
in: Approximation theory II, Academic Press 1976, 391-396

[62] R.Holmes-B.Kripke: Best approximation by compact operators
Indiana Univ. Math. J. 21, 1971, 255-263

[63] R.Holmes-B.Scranton-J.Ward: Approximation from the space of compact operators and other M-ideals
Duke Math. J. 42, 1975, 259-269

[64] R.Holmes-B.Scranton-J.Ward: Best approximation by compact operators II
Bull. of the Am. Math. Soc. 80, 1974,98-102

[65] D.H.Hyers-S.M.Ulam: Approximate isometries of the space of continuous functions
Ann. of Math. 48, 1947, 285-289

[66] M.Jerison: The space of bounded maps into a Banach space
Ann. of Math. 52, 1950, 309-327

[67] R.V.Kadison: Isometries of operator algebras
Ann. of Math. 54, 1951, 325-338

[68] I.Kaplansky: The structure of certain operator algebras
Trans. of the Am. Math. Soc. 70, 1951, 229-255

[69] J.Ketonen: The structure of countable Boolean algebras
Ann. of Math. 108, 1978, 41-89

[70] A.Lima: Intersection properteis of balls and subspaces in Banach spaces
Trans. of the Am. Math. Soc. 227, 1977, 1-62

[71] S.Machado-J.B.Prolla: An introduction to Nachbin spaces
Rend. del circ. Mat. di Palermo 21, 1972 119-139

[72] M.Meyer: Richesse du centre d'un espace vectorial réticulé
C. R. 283, 1976, 839-841

[73] E.Michael: Continuous selections I
Ann. of Math. 63, 1956, 361-382

[74] S.B.Myers: Banach spaces of continuous functions
Ann. of Math. 49, 1948, 132-140

[75] L.Nachbin-S.Machado-J.B.Prolla: Weighted approximation, vector fibrations and algebras of operators
J. Math. pures et appl. 50, 1971, 299-323

[76] J.v.Neumann: On rings or operators (reduction theory)
Ann. of Math. 50, 1949, 401-485

[77] N.M.Roy: A characterization of square Banach spaces
Israel J. of Math. 17, 1974, 142-148

[78] N.M.Roy: Contractive projections in square Banach spaces
Proc. of the Am. Math. Soc. 59, 1976, 291-296

[79] K.Saatkamp: M-ideals of compact operators
Math. Zeitschrift 158, 1978, 253-263

[80] U.Schirmeier: Ein Banach-Stone-Satz für adaptierte Vektorverbände und Algebren
Math. Zeitschrift 156, 1977, 279-290

[81] Z.Semadeni: Banach spaces of continuous functions I
Warschau, 1971

[82] R.R.Smith-J.D.Ward: M-ideal structure in Banach algebras
J. of Funct. Anal. 27, 1978, 337-349

[83] M.H.Stone: Applications of the theory of Boolean rings to general topology
Trans. of the Am. Math. Soc. 41, 1937, 375-481

[84] K.Sundaresan: Spaces of continuous functions into a Banach space
Studia Math. 48, 1973, 15-22

[85] F.E.Sullivan: Structure of real L^p-spaces
J. of Math. Anal. and Appl. 32, 1970, 621-629

[86] J.Varela: Sectional representation of Banach modules
Math. Zeitschrift 139, 1974, 55-61

[87] R.Walker: The Stone-Čech compactification
Springer Verlag, Erg. d. Math. 83, Berlin 1974

[88] A.W.Wickstead: The centraliser of $E \otimes_{\lambda} F$
Pac. J. of Math. 65, 1976, 563-571

[89] W.Wils: The ideal center of partially ordered vector spaces
Acta Math. 127, 1971, 41-77

Vol. 640: J. L. Dupont, Curvature and Characteristic Classes. X, 175 pages. 1978.

Vol. 641: Séminaire d'Algèbre Paul Dubreil, Proceedings Paris 1976–1977. Edité par M. P. Malliavin. IV, 367 pages. 1978.

Vol. 642: Theory and Applications of Graphs, Proceedings, Michigan 1976. Edited by Y. Alavi and D. R. Lick. XIV, 635 pages. 1978.

Vol. 643: M. Davis, Multiaxial Actions on Manifolds. VI, 141 pages. 1978.

Vol. 644: Vector Space Measures and Applications I, Proceedings 1977. Edited by R. M. Aron and S. Dineen. VIII, 451 pages. 1978.

Vol. 645: Vector Space Measures and Applications II, Proceedings 1977. Edited by R. M. Aron and S. Dineen. VIII, 218 pages. 1978.

Vol. 646: O. Tammi, Extremum Problems for Bounded Univalent Functions. VIII, 313 pages. 1978.

Vol. 647: L. J. Ratliff, Jr., Chain Conjectures in Ring Theory. VIII, 133 pages. 1978.

Vol. 648: Nonlinear Partial Differential Equations and Applications, Proceedings, Indiana 1976–1977. Edited by J. M. Chadam. VI, 206 pages. 1978.

Vol. 649: Séminaire de Probabilités XII, Proceedings, Strasbourg, 1976–1977. Edité par C. Dellacherie, P. A. Meyer et M. Weil. VIII, 805 pages. 1978.

Vol. 650: C*-Algebras and Applications to Physics. Proceedings 1977. Edited by H. Araki and R. V. Kadison. V, 192 pages. 1978.

Vol. 651: P. W. Michor, Functors and Categories of Banach Spaces. VI, 99 pages. 1978.

Vol. 652: Differential Topology, Foliations and Gelfand-Fuks-Cohomology, Proceedings 1976. Edited by P. A. Schweitzer. XIV, 252 pages. 1978.

Vol. 653: Locally Interacting Systems and Their Application in Biology. Proceedings, 1976. Edited by R. L. Dobrushin, V. I. Kryukov and A. L. Toom. XI, 202 pages. 1978.

Vol. 654: J. P. Buhler, Icosahedral Golois Representations. III, 143 pages. 1978.

Vol. 655: R. Baeza, Quadratic Forms Over Semilocal Rings. VI, 199 pages. 1978.

Vol. 656: Probability Theory on Vector Spaces. Proceedings, 1977. Edited by A. Weron. VIII, 274 pages. 1978.

Vol. 657: Geometric Applications of Homotopy Theory I, Proceedings 1977. Edited by M. G. Barratt and M. E. Mahowald. VIII, 459 pages. 1978.

Vol. 658: Geometric Applications of Homotopy Theory II, Proceedings 1977. Edited by M. G. Barratt and M. E. Mahowald. VIII, 487 pages. 1978.

Vol. 659: Bruckner, Differentiation of Real Functions. X, 247 pages. 1978.

Vol. 660: Equations aux Dérivée Partielles. Proceedings, 1977. Edité par Pham The Lai. VI, 216 pages. 1978.

Vol. 661: P. T. Johnstone, R. Paré, R. D. Rosebrugh, D. Schumacher, R. J. Wood, and G. C. Wraith, Indexed Categories and Their Applications. VII, 260 pages. 1978.

Vol. 662: Akin, The Metric Theory of Banach Manifolds. XIX, 306 pages. 1978.

Vol. 663: J. F. Berglund, H. D. Junghenn, P. Milnes, Compact Right Topological Semigroups and Generalizations of Almost Periodicity. X, 243 pages. 1978.

Vol. 664: Algebraic and Geometric Topology, Proceedings, 1977. Edited by K. C. Millett. XI, 240 pages. 1978.

Vol. 665: Journées d'Analyse Non Linéaire. Proceedings, 1977. Edité par P. Bénilan et J. Robert. VIII, 256 pages. 1978.

Vol. 666: B. Beauzamy, Espaces d'Interpolation Réels: Topologie et Géometrie. X, 104 pages. 1978.

Vol. 667: J. Gilewicz, Approximants de Padé. XIV, 511 pages. 1978.

Vol. 668: The Structure of Attractors in Dynamical Systems. Proceedings, 1977. Edited by J. C. Martin, N. G. Markley and W. Perrizo. VI, 264 pages. 1978.

Vol. 669: Higher Set Theory. Proceedings, 1977. Edited by G. H. Müller and D. S. Scott. XII, 476 pages. 1978.

Vol. 670: Fonctions de Plusieurs Variables Complexes III, Proceedings, 1977. Edité par F. Norguet. XII, 394 pages. 1978.

Vol. 671: R. T. Smythe and J. C. Wierman, First-Passage Perculation on the Square Lattice. VIII, 196 pages. 1978.

Vol. 672: R. L. Taylor, Stochastic Convergence of Weighted Sums of Random Elements in Linear Spaces. VII, 216 pages. 1978.

Vol. 673: Algebraic Topology, Proceedings 1977. Edited by P. Hoffman, R. Piccinini and D. Sjerve. VI, 278 pages. 1978.

Vol. 674: Z. Fiedorowicz and S. Priddy, Homology of Classical Groups Over Finite Fields and Their Associated Infinite Loop Spaces. VI, 434 pages. 1978.

Vol. 675: J. Galambos and S. Kotz, Characterizations of Probability Distributions. VIII, 169 pages. 1978.

Vol. 676: Differential Geometrical Methods in Mathematical Physics II, Proceedings, 1977. Edited by K. Bleuler, H. R. Petry and A. Reetz. VI, 626 pages. 1978.

Vol. 677: Séminaire Bourbaki, vol. 1976/77, Exposés 489–506. IV, 264 pages. 1978.

Vol. 678: D. Dacunha-Castelle, H. Heyer et B. Roynette. Ecole d'Eté de Probabilités de Saint-Flour. VII-1977. Edité par P. L. Hennequin. IX, 379 pages. 1978.

Vol. 679: Numerical Treatment of Differential Equations in Applications, Proceedings, 1977. Edited by R. Ansorge and W. Törnig. IX, 163 pages. 1978.

Vol. 680: Mathematical Control Theory, Proceedings, 1977. Edited by W. A. Coppel. IX, 257 pages. 1978.

Vol. 681: Séminaire de Théorie du Potentiel Paris, No. 3, Directeurs: M. Brelot, G. Choquet et J. Deny. Rédacteurs: F. Hirsch et G. Mokobodzki. VII, 294 pages. 1978.

Vol. 682: G. D. James, The Representation Theory of the Symmetric Groups. V, 156 pages. 1978.

Vol. 683: Variétés Analytiques Compactes, Proceedings, 1977. Edité par Y. Hervier et A. Hirschowitz. V, 248 pages. 1978.

Vol. 684: E. E. Rosinger, Distributions and Nonlinear Partial Differential Equations. XI, 146 pages. 1978.

Vol. 685: Knot Theory, Proceedings, 1977. Edited by J. C. Hausmann. VII, 311 pages. 1978.

Vol. 686: Combinatorial Mathematics, Proceedings, 1977. Edited by D. A. Holton and J. Seberry. IX, 353 pages. 1978.

Vol. 687: Algebraic Geometry, Proceedings, 1977. Edited by L. D. Olson. V, 244 pages. 1978.

Vol. 688: J. Dydak and J. Segal, Shape Theory. VI, 150 pages. 1978.

Vol. 689: Cabal Seminar 76–77, Proceedings, 1976–77. Edited by A.S. Kechris and Y. N. Moschovakis. V, 282 pages. 1978.

Vol. 690: W. J. J. Rey, Robust Statistical Methods. VI, 128 pages. 1978.

Vol. 691: G. Viennot, Algèbres de Lie Libres et Monoïdes Libres. III, 124 pages. 1978.

Vol. 692: T. Husain and S. M. Khaleelulla, Barrelledness in Topological and Ordered Vector Spaces. IX, 258 pages. 1978.

Vol. 693: Hilbert Space Operators, Proceedings, 1977. Edited by J. M. Bachar Jr. and D. W. Hadwin. VIII, 184 pages. 1978.

Vol. 694: Séminaire Pierre Lelong – Henri Skoda (Analyse) Année 1976/77. VII, 334 pages. 1978.

Vol. 695: Measure Theory Applications to Stochastic Analysis, Proceedings, 1977. Edited by G. Kallianpur and D. Kölzow. XII, 261 pages. 1978.

Vol. 696: P. J. Feinsilver, Special Functions, Probability Semigroups, and Hamiltonian Flows. VI, 112 pages. 1978.

Vol. 697: Topics in Algebra, Proceedings, 1978. Edited by M. F. Newman. XI, 229 pages. 1978.

Vol. 698: E. Grosswald, Bessel Polynomials. XIV, 182 pages. 1978.

Vol. 699: R. E. Greene and H.-H. Wu, Function Theory on Manifolds Which Possess a Pole. III, 215 pages. 1979.